职业教育示范性规划教材

网络管理与维护项目实训教程

主　编　王　苒　　邱建伟

副主编　王晓亮

主　审　王　博

电子工业出版社

Publishing House of Electronics Industry

北京·BEIJING

内 容 简 介

本书以网络管理与维护的认知流程为脉络,详细讲述了网络设备的各项常用配置以及相关应用的配置。本书主要内容包括:交换机的各项基础配置和具体应用配置方法,路由器的各项基础配置和具体应用配置方法,防火墙的各项基础配置和具体应用配置方法等。所有实训均以现实网络环境和应用为背景,按照学习过程逐步展开,方便学生学习和掌握。

本书编写过程中参考了国家和行业的相关认证的标准要求,适合作为职业院校信息技术类专业的网络设备课程实训指导用书,也可以作为从事系统集成工作的技术人员参考用书。

未经许可,不得以任何方式复制或抄袭本书之部分或全部内容。
版权所有,侵权必究。

图书在版编目(CIP)数据

网络管理与维护项目实训教程 / 王苒,邱建伟主编. —北京:电子工业出版社,2014.6
职业教育示范性规划教材
ISBN 978-7-121-23317-3

Ⅰ. ①网… Ⅱ. ①王… ②邱… Ⅲ. ①计算机网络管理-中等专业学校-教材②计算机网络-计算机维护-中等专业学校-教材 Ⅳ. ①TP393.07

中国版本图书馆 CIP 数据核字(2014)第 110403 号

策划编辑:白　楠
责任编辑:白　楠
印　　刷:北京虎彩文化传播有限公司
装　　订:北京虎彩文化传播有限公司
出版发行:电子工业出版社
　　　　　北京市海淀区万寿路 173 信箱　邮编　100036
开　　本:787×1 092　1/16　印张:11.5　字数:294.4 千字
版　　次:2014 年 6 月第 1 版
印　　次:2018 年 8 月第 3 次印刷
定　　价:25.00 元

凡所购买电子工业出版社图书有缺损问题,请向购买书店调换。若书店售缺,请与本社发行部联系,联系及邮购电话:(010)88254888,88258888。
质量投诉请发邮件至 zlts@phei.com.cn,盗版侵权举报请发邮件至 dbqq@phei.com.cn。
本书咨询联系方式:(010)88254592,bain@phei.com.cn。

前　言

随着计算机网络与通信技术的发展，计算机网络在人们的生活、学习和工作中的位置越来越重要，网络的管理和维护已经成为确保网络使用率的重要环节。因此，能够掌握一定的网络设备的配置和管理已成为计算机网络行业从业人员的一种必需技能。

本书介绍了常见的网络设备的基础配置和常见应用的配置方式方法，针对各种配置的用途也做了简要的介绍。本书以实际网络环境为背景，以实际网络应用为案例，分析了所需的配置方式和配置命令。使得学习者不但能配置相关的网络应用，还能掌握配置相关应用的思想和思维方式，了解从应用到配置的整个过程，以适应现代网络管理的需要。

本书共 7 个项目，具体内容如下：

项目 1 着重讲述了交换机的基础配置和相关配置命令，以及交换机的基础知识。

项目 2 重点讲述了二层交换机的相关配置以及相应网络应用的配置方式方法和所需的配置命令。

项目 3 重点讲述了三层交换机的相关配置以及相应网络应用的配置方式方法和所需的配置命令。

项目 4 着重讲述了路由器的基础配置和相关配置命令，以及路由器的基础知识。

项目 5 重点讲述了路由器的常见应用和相应的配置方式方法。

项目 6 着重讲述了防火墙的基础配置和相关配置命令。

项目 7 重点讲述了防火墙的必需配置以及相应网络安全应用的配置方式、方法和所需的配置命令。

本书由王苒、邱建伟主编，王晓亮副主编，神州数码有限公司工程师王博主审。编写时还采纳了许多系统集成一线工程师的意见和建议，并参考了国家和行业的相关认证的标准要求。

本书内容翔实，全部采用实例教学，并配合大量的实训任务，让读者在不断的操作过程中掌握计算机网络管理与维护技能。

在本书编写过程中，虽然作者倾注了大量的心血，但是仍然可能存在疏漏，还请广大读者不吝赐教。

编者

目　　录

项目 1　交换机的管理和维护 ··· 1
　　1.1　实训一　认识交换机 ··· 2
　　1.2　实训二　交换机带外及带内管理 ··· 3
　　1.3　实训三　交换机的配置模式 ·· 5
　　1.4　实训四　交换机的端口名称配置 ··· 8
　　1.5　实训五　设置交换机 IP 地址 ·· 9
　　1.6　实训六　交换机账号的管理 ·· 11
　　1.7　实训七　管理交换机系统文件 ·· 12
　　1.8　实训八　恢复交换机出厂配置与保存当前配置 ·· 14

项目 2　二层交换机的基础应用 ··· 16
　　2.1　实训一　交换机端口配置 ··· 17
　　2.2　实训二　单台交换机 VLAN 划分 ··· 19
　　2.3　实训三　跨交换机相同 VLAN 访问 ·· 22
　　2.4　实训四　交换机私有 VLAN 的应用 ·· 28
　　2.5　实训五　提升交换机间的连接带宽 ·· 31
　　2.6　实训六　交换机端口镜像 ··· 34
　　2.7　实训七　避免网络的冗余链路危害 ·· 35

项目 3　三层交换机的应用 ·· 42
　　3.1　实训一　三层交换机 VLAN 的划分与 VLAN 间路由 ································ 43
　　3.2　实训二　使用三层交换机实现二层交换机 VLAN 间的路由 ······················· 46
　　3.3　实训三　三层交换机静态路由配置 ·· 53
　　3.4　实训四　三层交换机 RIP 动态路由配置 ·· 61
　　3.5　实训五　三层交换机 OSPF 动态路由配置 ··· 69
　　3.6　实训六　三层交换机 ACL 访问控制列表配置 ·· 77
　　3.7　实训七　三层交换机 MAC 与 IP 绑定 ··· 79
　　3.8　实训八　三层交换机配置 DHCP 服务器 ··· 82
　　3.9　实训九　三层交换机 DHCP 中继配置 ··· 85

V

项目 4 路由器的管理和维护 ………………………………………………… 90
4.1 实训一 认识路由器模块和端口 …………………………………… 91
4.2 实训二 路由器带外及带内管理 …………………………………… 92
4.3 实训三 路由器配置模式 ………………………………………… 94
4.4 实训四 管理路由器账号 ………………………………………… 96
4.5 实训五 管理路由器系统文件 …………………………………… 97

项目 5 路由器的基础应用 ………………………………………………… 100
5.1 实训一 路由器单臂路由配置 …………………………………… 101
5.2 实训二 路由器静态路由配置 …………………………………… 103
5.3 实训三 路由器 RIP 协议配置 …………………………………… 108
5.4 实训四 路由器 OSPF 配置 …………………………………… 114
5.5 实训五 路由器串口 PPP 封装 …………………………………… 119
5.6 实训六 PPP 封装协议 CHAP 认证 …………………………………… 122
5.7 实训七 NAT 网络地址转换 …………………………………… 127

项目 6 认识防火墙 ………………………………………………… 130
6.1 实训一 防火墙外观与端口 …………………………………… 131
6.2 实训二 防火墙管理模式 …………………………………… 133
6.3 实训三 管理防火墙配置文件 …………………………………… 136

项目 7 防火墙的具体应用 ………………………………………………… 138
7.1 实训一 配置防火墙网络接口 IP 及 zone …………………………………… 139
7.2 实训二 配置防火墙 DNAT …………………………………… 143
7.3 实训三 配置防火墙 DHCP …………………………………… 146
7.4 实训四 配置防火墙 URL 过滤 …………………………………… 148
7.5 实训五 配置防火墙网页内容过滤 …………………………………… 152
7.6 实训六 防火墙 Web 认证配置 …………………………………… 157
7.7 实训七 防火墙 IP-MAC 绑定配置 …………………………………… 164
7.8 实训八 防火墙负载均衡配置 …………………………………… 165
7.9 实训九 防火墙禁用实时通信类工具配置 …………………………………… 168
7.10 实训十 防火墙应用 QOS 配置 …………………………………… 170
7.11 实训十一 配置防火墙 SSL VPN …………………………………… 173

项目 1

交换机的管理和维护

教学目标

通过本章的学习,学生可以了解交换机的基础配置命令以及基础配置的方式和方法,为接下来的交换机具体应用打好基础。

能力目标

了解交换机在网络中的作用
熟悉交换机的基础配置
掌握交换机的基础配置命令

知识目标

熟悉交换机的端口分类
熟悉交换机的配置模式
熟悉交换机的配置方法

主要教学内容

交换机的管理
交换机的物理和逻辑端口
交换机的IP地址配置

1.1 实训一 认识交换机

要想正确地配置交换机,首先需要认识交换机。本书是一本实训指导类教材,所以对交换机在网络中的具体作用不再赘述。

任务 认识交换机的模块和端口,熟悉端口的属性以及指示灯的状态含义。

任务准备

实验所需设备为 DCS-3950-28C 交换机一台。

任务实施

DCS-3950-28C 交换机的前面板上有 24 个 10/100Base-T 端口,4 个 Combo 端口(4 个 RJ-45 和 4 个 SFP 端口),1 个 Console 端口和 30 个 LED 指示灯。后面板包括 1 个 220V 交流电源插座。

1. DCS-3950-28C 交换机的前面板(如图 1-1 所示)。

图 1-1 交换机前面板

2. 指示灯说明。

电源和连接指示灯的状态及含义见表 1-1。

表1-1 电源和连接指示灯的状态及含义

LED	状态	含义
Power	亮(绿色)	内部电源工作正常
	灭	电源没上电或电源坏
DIAG	闪(绿色 1Hz)	运行状态正常
	闪(绿色 8Hz)	系统加载中

端口指示灯状态及含义见表 1-2。

表1-2 端口指示灯状态及含义

面板标记	状态	含义
Port1-24(Link/Act)	亮(绿色)	端口 link 成功
	闪(绿色)	端口 link 成功,并收发数据
	灭	端口没有 link 成功
Port 25/26/27/28(Link/Act,1000M)	亮(绿色)	Combo 端口 link 成功
	闪(绿色)	Combo 端口 link 成功,并收发数据
	灭	Combo 端口没有 link 成功

各端口属性见表 1-3。

表1-3 端口属性说明

端口形式	规　　格
RJ-45 端口	● 10/100Mbps 自适应 ● MDI/MDI-X 网线类型自适应 ● 5 类非屏蔽双绞线（UTP）：100m ● Combo 端口支持 10/100/1000Mbps 自适应
SFP	● SFP-SX-L 收发器： 1000Base-SX SFP（850nm, MMF, 550m） ● SFP-LX-L 收发器： 1000Base-LX SFP 接口卡模块（1310nm, SMF, 10km 或 MMF, 550m） ● SFP-LX-20-L 收发器： 1310nm 光波，9/125 μm 单模光纤：20km ● SFP-LX-40 收发器： 9/125μm 单模光纤：40km ● SFP-LH-70-L 收发器： 9/125μm 单模光纤：70 km ● SFP-LH-120-L 收发器： 9/125μm 单模光纤：120 km

1.2　实训二　交换机带外及带内管理

任务 了解交换机的带外及带内管理，并掌握交换机最基本的管理方式——带外管理的方法。

任务描述

网络设备的管理方式可以简单地分为带外管理（out-of-band）和带内管理（in-band）两种管理模式。所谓带内管理，是指网络的管理控制信息与用户网络的承载业务信息通过同一个逻辑信道传送（例如 telnet 登录），需要占用业务带宽；而在带外管理模式中，网络的管理控制信息与用户网络的承载业务信息在不同的逻辑信道传送，也就是设备提供专门用于管理的带宽。目前，很多高端的交换机都有带外网管接口，使网络管理的带宽和业务带宽完全隔离，互不影响，构成单独的管理网络。

通过 Console 端口管理是最常用的带外管理方式，通常用户会在首次配置交换机或者无法进行带内管理时使用带外管理方式。带外管理方式也是使用频率最高的管理方式。带外管理的时候，使用 Windows 操作系统自带的"超级终端"程序来连接交换机。Console 端口也叫配置口，是符合 RS-232 标准的接口，用于连接交换机。交换机包装箱中标配该端口线缆，用于连接 Console 端口和配置终端。

任务准备

实验所需设备为 DCS-3950-28C 交换机一台，PC 一台，配置线缆一根。实验拓扑图如图 1-2 所示。

图 1-2　实验拓扑图

任务实施

1. 连接配置线缆。

拔插 Console 线时注意保护交换机的 Console 端口和 PC 的串口，不要带电拔插。

2. 使用超级终端连接交换机。

在 Windows 操作系统中，依次单击"开始"→"程序"→"附件"→"通讯"→"超级终端"。

为建立的超级终端连接命名，并选择图标。这里我们用交换机的型号"DCS-3950-28C"作为名字，如图 1-3 所示。

选择使用的 COM 端口号。需要知道配置线连接的是计算机的哪个 COM 口，一般为"COM1"，如图 1-4 所示。

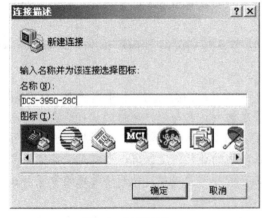
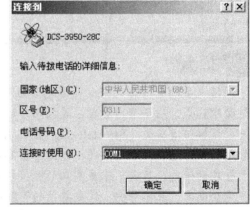

图 1-3　超级终端　　　　　　　　　图 1-4　选择连接端口

设置"COM1"端口属性，波特率为 9600，数据位 8，奇偶校验为"无"，停止位为"1"，数据流控制为"无"。也可以单击右下方的"还原为默认值"按扭来一键设置，如图 1-5 所示。

3．单击"确认"按钮，进入超级终端。按下"Enter"键，将会看到如图1-6所示界面，表示已经成功进入交换机，已经可以对交换机进行配置。

图1-5 设置连接参数

图1-6 交换机登入界面

接下来可以使用"show version"命令，查看交换机的软硬件版本信息，如图1-7所示。

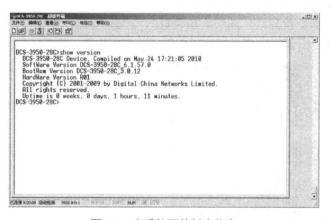

图1-7 查看软硬件版本信息

1.3　实训三　交换机的配置模式

认识了交换机的端口和属性，了解了使用带外管理来配置交换机的方法，接下来我们要掌握交换机的各种配置模式。

> **任务**
> 1．熟悉交换机不同配置模式的功能。
> 2．掌握交换机四种基本配置模式的进入命令和可选命令。

任务准备

实验所需设备为 DCS-3950-28C 交换机一台，PC 一台，配置线缆一根。实验拓扑图如图 1-8 所示。

图 1-8　实验拓扑图

任务实施

逐一认识交换机的四种配置模式

1. setup 配置模式。

交换机出厂后第一次启动，超级终端会进入 setup 视图。也可以通过"setup"命令进入该视图，如图 1-9 所示。

图 1-9　配置视图

依次输入 y，进入配置视图；输入 1，选择语言模式为 Chinese。

还可以通过输入选项数字，来配置其他信息。

[0]: 配置交换机主机名。

[1]: 配置 Vlan1 的接口。

[2]: 配置交换机 Telnet 服务器。

[3]: 配置交换机 Web 服务器。

[4]: 配置 SNMP。

[5]: 退出 setup 模式不保存配置结果。

[6]: 退出 setup 模式保存配置结果。

2. 一般用户配置模式。

退出 setup 视图或者建立好超级终端启动交换机后就会进入一般用户视图，提示符为 ">"。在这个模式中，我们可以进行切换语言模式等一些基本操作。其常用命令如图 1-10 所示。

3. 特权用户配置模式。

在一般用户视图输入"enable"命令就会进入特权用户视图，提示符为"#"，在这里我们可以查看交换机当前所有的有效配置，其常用命令如图 1-11 所示。

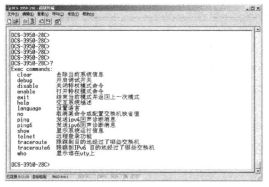

图 1-10　一般用户配置模式

图 1-11　特权用户配置模式

4. 全局配置模式。

在特权用户视图输入"config terminal"命令就会进入全局配置视图，提示符为"(config)#"，在这里我们可以完成所有对交换机的配置。其常用命令如图 1-12 所示。

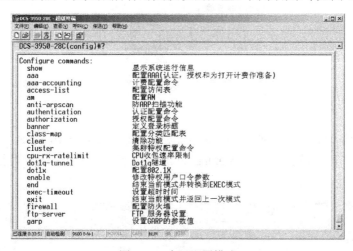

图 1-12　全局配置模式

以上我们所看到的界面视图都称为 CLI（Command Line Interface）界面，该界面也被称为命令行界面，它由 shell 程序提供，由一系列配置命令组成。每个界面视图的退出命令都是"exit"。

1.4 实训四 交换机的端口名称配置

> **任务** 掌握交换机各物理端口和配置名称的对应关系。

任务描述

在实际使用中，我们需要知道交换机各物理端口的配置名称，从而实现相应的配置操作。

任务准备

实验所需设备为 DCS-3950-28C 交换机一台，PC 一台，配置线缆一根，直通缆一根。实验拓扑图如图 1-13 所示。

图 1-13 实验拓扑图

任务实施

通过前面的学习，我们知道 DCS-3950-28C 交换机有 24 个 10/100Base-T 端口和 4 个 Combo 端口，我们可以通过 "show running-config" 命令来查看它们的配置名称，如图 1-14、图 1-15 所示。

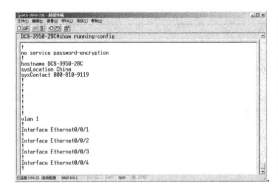

图 1-14 查看端口命令

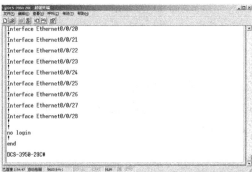

图 1-15 端口显示

可以看到，端口的配置名称为 0/0/端口号。以 0/0/1 为例，第一个 0 表示堆叠中的第一台交换机，如果是 1，就表示第 2 台交换机；第 2 个 0 表示交换机上的第 1 个模块；最后的 1 表示当前模块上的第 1 个网络端口。0/0/1 表示用户使用的是堆叠中第一台交换机网络端口模块上的第一个网络端口。默认情况下，如果不存在堆叠，交换机总会认为自己是第 0 台交换机。进入端口的命令为："DCS-3950-28C(config)#interface ethernet 0/0/1"。可以使用"？"命令查看可用的命令，如图 1-16、图 1-17 所示。

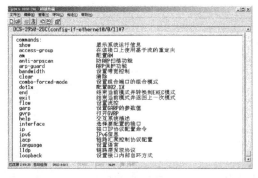
图 1-16　查看端口配置可用命令 1

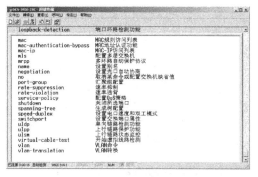
图 1-17　查看端口配置可用命令 2

1.5　实训五　设置交换机 IP 地址

任务　通过给交换机设置 IP 地址来实现 Telnet 登录交换机。

任务描述

前面，我们学习了交换机的各种配置界面，登录这些界面都需要配置线缆连接计算机和交换机的 Console 端口。而在实际应用过程中，管理员很难做到带着笔记本电脑、配置线缆去不同的位置调试每台交换机。实际使用中，我们是借助网线以 Telnet 的方式来登录和管理交换机的。

任务准备

实验所需设备为 DCS-3950-28C 交换机一台，PC 一台，配置线缆一根，直通缆一根。实验拓扑图如图 1-18 所示。

图 1-18　实验拓扑图

任务实施

用直通缆连接 PC 机网卡和交换机的 16 端口。

1. 给交换机设置 IP 地址，命令如下：

```
DCS-3950-28C#config
DCS-3950-28C (Config)#interface vlan 1              //进入Vlan1接口
DCS-3950-28C(config-if-vlan1)#%Jan 01 01:07:08 2006 %LINK-5-CHANGED:
Interface Vlan1, changed state to UP                //Vlan1 启用
DCS-3950-28C (Config-If-Vlan1)#ip address 192.168.1.11 255.255.255.0
   //给Vlan1配置IP地址
DCS-3950-28C (Config-If-Vlan1)#no shutdown          //激活Vlan1接口
DCS-3950-28C#ping 192.168.1.10                      //测试和PC是否连通
DCS-3950-28C#ping 192.168.1.10
Type ^c to abort.
Sending 5 56-byte ICMP Echos to 192.168.1.10, timeout is 2 seconds.
!!!!!
Success rate is 100 percent (5/5), round-trip min/avg/max = 0/0/0 ms
   //出现5个！证明ping通
```

2. 交换机开启 Telnet 服务（系统默认打开 Telnet 服务器功能）。

```
DCS-3950-28C(config)#telnet-server enable
```

3. 为交换机设置授权 Telnet 用户。

```
DCS-3950-28C(config)#username test privilege 15 password 0 test    //
设置用户名 test，配置优先级 15（0～15，15的优先级最高），配置明文密码为 test（0
是明文密码，7是加密密码并且密码长度必须为32位）
DCS-3950-28C(config)#authentication line vty login local   //设置vty验证
方式为本地
```

4. 配置主机的 IP 地址为 192.168.1.10，在本实验中要与交换机的 IP 地址在一个网段。

5. 验证使用 Telnet 登录。输入正确的用户名和密码后，登录成功，如图 1-19 所示。

图 1-19 验证 Telnet 登录

6. 实验小结，默认情况下，交换机所有端口都属于 Vlan1，因此通常把 Vlan1 作为交换机的管理 Vlan，Vlan1 接口的 IP 地址就是交换机的管理地址。删除一个 Telnet 用户可以在 config 模式下使用"no username 用户名"命令。二层交换机，一般只支持一个 IP 地址用于 Telnet 管理。三层交换机能支持最多 4096 个 IP 地址，用来实现路由功能。

1.6　实训六　交换机账号的管理

任务 为交换机设置登录密码。

任务描述

为了防止有人登录交换机并修改配置文件，网络管理员一般都需要为交换机设置管理密码。

任务准备

实验所需设备为 DCS-3950-28C 交换机一台，PC 一台，配置线缆一根。实验拓扑图如图 1-20 所示。

图 1-20　实验拓扑图

任务实施

在全局配置模式下设置特权用户口令：

```
DCS-3950-28C >enable
DCS-3950-28C #config terminal                    //进入全局配置模式
DCS-3950-28C (Config)#enable password 0 admin    //设置特权配置模式密码为admin
DCS-3950-28C #write                              //保存配置
验证实验：
DCS-3950-28C #exit                               //退出特权用户配置模式
DCS-3950-28C>enable                              //进入特权模式
Password:                                        //提示输入登录密码
DCS-3950-28C#                                    //密码验证通过，进入特权模式
```

也可以通过验证方法 2："show"命令来查看我们的配置，如图 1-21 所示。

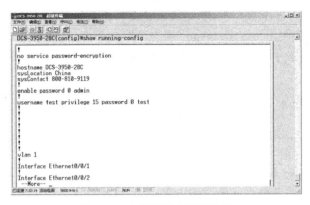

图 1-21　交换机账户配置验证

1.7　实训七　管理交换机系统文件

任 务　备份交换机系统文件。

任务描述

当我们完成对交换机的配置以后，接下来的工作就是把配置文件和系统文件从交换机里复制出来另外存放，防止意外情况的发生。当交换机出现故障以后，我们可以直接把备份的文件下载到交换机上，让网络能继续正常运作。交换机文件的备份需要采用TFTP 服务器。TFTP（Trivial File Transfer Protocol）主要用于主机之间、主机与交换机之间传输文件。它们都采用客户机-服务器模式进行文件传输。TFTP 相对于 FTP 的优点是提供简单的、开销不大的文件传输服务，更加适合交换机备份文件使用。我们使用tftpd32 作为 TFTP 服务器。

任务准备

实验所需设备为 DCS-3950-28C 交换机一台，PC 一台，配置线缆一根，直通缆一根。实验拓扑图如图 1-22 所示。

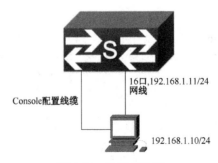

图 1-22　实验拓扑图

任务实施

实验步骤：

1. 配置 TFTP 服务器。

双击 tftpd32.exe，出现 TFTP 服务器的主界面，如图 1-23 所示。

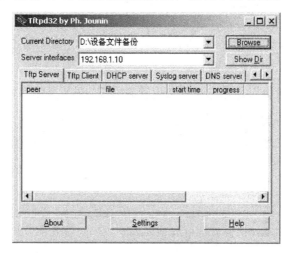

图 1-23　TFTP 服务器的主界面

在主界面中，我们看到该服务器的根目录是"D:\设备文件备份"，服务器的 IP 地址也自动出现在第二行：192.168.1.10。可以更改当前目录：单击"Browse"按钮进行设置，单击"确定"按钮进行保存确认。也可以单击"Settings"按钮进行其他设置。到此，TFTP 服务器搭建好了，可以将它最小化到右下角的工具栏中。

2. 给交换机设置 IP 地址即管理 IP。验证主机与交换机是否连通。

3. 查看需要备份的文件。

```
DCS-3950-28C#show flash
    nos.img                        4,623,669              //系统文件
    startup-config                 919                    //配置文件，需要保存
    Used   4,624,588 bytes in 2 files, Free  12,152,628 bytes.file name file
length
```

4. 备份配置文件。

```
DCS-3950-28C#copy startup-config tftp://192.168.1.10/startup        //复制
statrup-config到192.168.1.10并重命名为startup
    Confirm [Y/N]:y                                        //输入y，确认
    Begin to send file, please wait...                     //开始上传文件
    File transfer complete.
    close tftp client.
    DCS-3950-28C#                                          //传送成功
```

5. 备份系统文件。

```
DCS-3950-28C#copy nos.img tftp://192.168.1.10/nos          //复制nos.img到
```

192.168.1.10并重命名为nos
```
    Confirm copy file [Y/N]:y                          //输入y，确认
    Begin to send file, please wait...                 //开始上传文件
    ######################################################################
##
    File transfer complete.
    close tftp client.
    DCS-3950-28C#                                      //传送成功
```

6. 验证实验。

在"D:\设备文件备份"目录中查看上传的文件和大小，如图 1-24 所示。

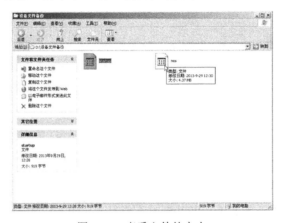

图 1-24　查看文件的大小

注意观察两个文件的大小，和交换机"show flash"查看到的文件是否一致。

1.8　实训八　恢复交换机出厂配置与保存当前配置

任务　恢复交换机出厂配置与保存当前配置。

任务描述

当做完实验需要重启交换机来验证实验效果时，需要使用保存命令。而当配置有问题时，需要将交换机恢复到出厂设置。

任务准备

实验所需设备为 DCS-3950-28C 交换机一台，PC 一台，配置线缆一根。实验拓扑图如图 1-25 所示。

任务实施

1. 保存交换机当前配置。将交换机名称修改为"switch"，并

图 1-25　实验拓扑图

保存，重启交换机验证。

```
DCS-3950-28C(config)#hostname switch          //将交换机改名为switch
switch(config)#                                //改名生效
switch#write                                   //在特权模式保存配置
switch#e%Jan 01 02:17:14 2006 Write configuration successfully!
switch#reload                                  //重启交换机
Process with reboot? [Y/N] y                   //确认重启
switch>                                        //提示符为switch
```

2. 恢复交换机到出厂设置。

前一个班级的同学完成了课程任务后，他们如果像刚才的实验一样，使用了"write"保存命令，会影响后续上课的其他班级，所以我们需要将交换机的所有配置信息清除，这时可以使用恢复交换机到出厂设置命令。

```
switch>enable                                  //进入特权用户配置模式
switch#set default                             //使用"set default"命令
Are you sure? [Y/N] = y                        //确认
switch#write                                   //保存配置
switch#%Jan 01 00:03:06 2006 Switch configuration has been set default!
switch#show startup-config                     //查看当前的配置文件
% Current startup-configuration is default factory configuration!
//提示是默认配置
switch#reload                                  //重启交换机
Process with reboot? [Y/N] y                   //确认重启
```

结果如图 1-26 所示，可以看到当前进入了 setup 界面，而且交换机的名称又变回了 DCS-3950-28C（注意，恢复出厂设置后，一定要用"write"命令）。

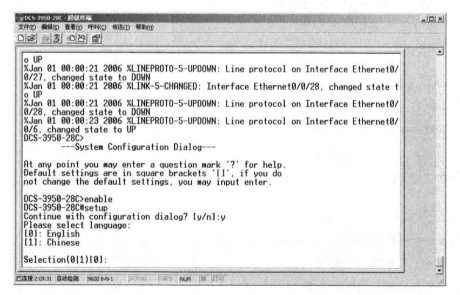

图 1-26 恢复出厂设置

项目 2

二层交换机的基础应用

教学目标

通过本章的学习，学生可以了解二层交换机的基础配置命令以及针对基础配置的方式和方法。掌握针对二层交换机虚拟局域网的应用配置。为接下来的三层交换机的具体应用配置打好基础。

能力目标
了解二层交换机的作用
熟悉二层交换机的基础配置
掌握虚拟局域网的配置

知识目标
熟悉虚拟局域网的概念
熟悉负载均衡的概念
熟悉网络监控的概念

主要教学内容
各种VLAN的配置
网络负载均衡的配置
交换机端口监听的配置

2.1 实训一 交换机端口配置

任务描述

某公司有两台二层交换机 DCS-3950，为了实现两台交换机间高速传输数据，连接这两台交换机的 25 端口，分别设置速度为 1000Mbps，双工模式为全双工，以及设定广播风暴控制。

本任务采用如下设备：DCS-3950 交换机两台，PC 一台，Console 线缆一根，直通网线一根。

网络拓扑图如图 2-1 所示。

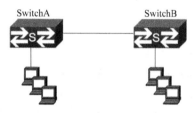

图 2-1 网络拓扑图

任务准备

半双工：半双工（Half Duplex）数据传输指的是数据可以在一个信号载体的两个方向上传输，但是不能同时传输。

全双工：全双工（Full Duplex）是在两个设备之间采用发送线路和接收线路各自独立的方法，这种方法可以使数据在两个方向上同时进行传送操作。也就是在发送数据的同时也能够接收数据，两者同步进行。

单工：单工（Simplex Communication）模式的数据传输是单向的。通信双方中，一方固定为发送端，另一方则固定为接收端。信息只能沿一个方向传输，使用一根传输线。

任务实施

SwitchA

1. 进入交换机端口配置模式。

```
DCS-3950-28C>enable
DCS-3950-28C#config
DCS-3950-28C(config)#hostname SwitchA          //设置交换机标识符为 SwitchA
SwitchA(config)#interface ethernet 1/25        //进入 E1/25 端口
```

2. 配置交换机端口速度和双工模式。

```
SwitchA(config-if-ethernet1/25)#speed-duplex force1g-full
                //设置端口速度为 1000Mbps，双工模式为全双工
```

3. 配置交换机端口风暴控制。

```
SwitchA(config-if-ethernet1/25)#storm-control broadcast 1000
```

```
                          //交换机每秒允许通过的广播1000个数据包
SwitchA(config-if-ethernet1/25)#storm-control unicast 1000
                          //交换机每秒允许通过的未知目的单播为1000个数据包
SwitchA(config-if-ethernet1/25)#exit       //退出
```

4．检测交换机配置。

```
SwitchA(config)#show running-config
no service password-encryption
hostname SwitchA                          //交换机标识符为SwitchA
sysLocation China
sysContact 800-810-9119
username admin privilege 15 password 0 admin
vlan 1
Interface Ethernet1/1
Interface Ethernet1/2
略
Interface Ethernet1/23
Interface Ethernet1/24
Interface Ethernet1/25
 speed-duplex force1g-full     //端口速度为1000Mbps，双工模式为全双工
 storm-control broadcast 1000  //每秒允许通过的广播1000个数据包
 storm-control unicast 1000    //每秒允许通过的未知目的的单播1000个数据包
Interface Ethernet1/26
Interface Ethernet1/27
Interface Ethernet1/28
interface Vlan1
 ip address 192.168.1.1 255.255.255.0     //交换机默认管理IP
no login
End
```

SwitchB

1．进入交换机端口配置模式。

```
DCS-3950-28C>enable
DCS-3950-28C#config
DCS-3950-28C(config)#hostname SwitchB     //设置交换机标识符为SwitchB
SwitchB(config)#interface ethernet 1/25   //进入E1/25端口
```

2．配置交换机端口速度和双工模式。

```
SwitchB(config-if-ethernet1/25)#speed-duplex force1g-full
                          //设置端口速度为1000Mbps，双工模式为全双工
```

3．配置交换机端口风暴控制。

```
SwitchB(config-if-ethernet1/25)#storm-control broadcast 1000
                          //交换机每秒允许通过的广播1000个数据包
SwitchB(config-if-ethernet1/25)#storm-control unicast 1000
                          //交换机每秒允许通过的未知目的的单播为1000个数据包
SwitchB(config-if-ethernet1/25)#exit       //退出
```

4. 检测交换机配置。

```
SwitchB(config)#show running-config
no service password-encryption
hostname SwitchB                              //交换机标识符为SwitchB
sysLocation China
sysContact 800-810-9119
username admin privilege 15 password 0 admin
vlan 1
Interface Ethernet1/1
Interface Ethernet1/2
略
Interface Ethernet1/25
 speed-duplex force1g-full         //端口速度为1000Mbps，双工模式为全双工
 storm-control broadcast 1000      //每秒允许通过的广播1000个数据包
 storm-control unicast 1000        //每秒允许通过的未知目的的单播1000个数据包
Interface Ethernet1/26
Interface Ethernet1/27
Interface Ethernet1/28
interface Vlan1
 ip address 192.168.1.1 255.255.2525.0         //交换机默认管理IP
no login
End
```

2.2 实训二 单台交换机 VLAN 划分

任务描述

某小学同一楼层有两个一体化教室，两个一体化教室的信息端口都连接在一台二层交换机上。学校已经为一体化教室分配了固定的 IP 地址段，为了保证两个一体化教室的相对独立，教室之间数据互不干扰。管理员设置两个 VLAN，并且交换机 1~12 端口属于 VLAN 100，13~24 端口属于 VLAN 200。

本任务采用如下设备：DCS-3950 交换机一台，PC 两台，Console 线缆一根，直通网线两根。设备 IP 地址分配见表 2-1。

表2-1　IP地址分配

设备名称	IP 地址	所属 VLAN	连接端口
PC1	192.168.1.101/24	VLAN100	E1/1
PC2	192.168.1.201/24	VLAN200	E1/13

网络拓扑图如图 2-2 所示。

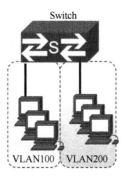

图 2-2　网络拓扑图

任务准备

VLAN：VLAN（Virtual Local Area Network）的中文名为"虚拟局域网"。VLAN 是一种将局域网设备从逻辑上划分成一个个网段，从而实现虚拟工作组的数据交换技术。

VLAN 是为解决以太网的广播问题和安全性而提出的一种协议，协议名为 IEEE 802.1Q，它在以太网帧的基础上增加了标签（tag），其中包含 VLAN ID 号，用 VLAN ID 把用户划分为不同的工作组，每个工作组就是一个虚拟局域网，也就从逻辑上限制不同工作组间的用户互访。一个 VLAN 内部的广播和单播流量都不会转发到其他 VLAN 中，从而有助于控制流量、减少设备投资、简化网络管理、提高网络的安全性。

任务实施

1. 创建 VLAN 100 并加入相应端口。

```
DCS-3950-28C>enable
DCS-3950-28C#config
DCS-3950-28C(config)#hostname Switch          //设置交换机名为 Switch
Switch(config)#vlan 100                        //创建 VLAN 100
Switch(config-vlan100)#switchport interface ethernet 1/1-12
//将 1~12 端口加入 VLAN 100
Set the port Ethernet1/1 access vlan 100 successfully
Set the port Ethernet1/2 access vlan 100 successfully
Set the port Ethernet1/3 access vlan 100 successfully
Set the port Ethernet1/4 access vlan 100 successfully
Set the port Ethernet1/5 access vlan 100 successfully
Set the port Ethernet1/6 access vlan 100 successfully
Set the port Ethernet1/7 access vlan 100 successfully
Set the port Ethernet1/8 access vlan 100 successfully
Set the port Ethernet1/9 access vlan 100 successfully
Set the port Ethernet1/10 access vlan 100 successfully
Set the port Ethernet1/11 access vlan 100 successfully
Set the port Ethernet1/12 access vlan 100 successfully
Switch(config-vlan100)#exit                    //退出
```

2. 创建 VLAN 200 并加入相应端口。

```
Switch(config)#vlan 200                         //创建VLAN 200
Switch(config-vlan200)#switchport interface ethernet 1/13-24
//将13～24端口加入VLAN 200
Set the port Ethernet1/13 access vlan 200 successfully
Set the port Ethernet1/14 access vlan 200 successfully
Set the port Ethernet1/15 access vlan 200 successfully
Set the port Ethernet1/16 access vlan 200 successfully
Set the port Ethernet1/17 access vlan 200 successfully
Set the port Ethernet1/18 access vlan 200 successfully
Set the port Ethernet1/19 access vlan 200 successfully
Set the port Ethernet1/20 access vlan 200 successfully
Set the port Ethernet1/21 access vlan 200 successfully
Set the port Ethernet1/22 access vlan 200 successfully
Set the port Ethernet1/23 access vlan 200 successfully
Set the port Ethernet1/24 access vlan 200 successfully
Switch(config-vlan200)#exit                     //退出
```

3. 检测配置。

```
Switch(config)#show running-config              //显示当前系统配置
no service password-encryption
hostname Switch
sysLocation China
sysContact 800-810-9119
username admin privilege 15 password 0 admin
vlan 1;100;200                                  //交换机已有VLAN
Interface Ethernet1/1
 switchport access vlan 100                     //端口E1/1属于VLAN100
Interface Ethernet1/2
 switchport access vlan 100                     //端口E1/2属于VLAN100
Interface Ethernet1/3
 switchport access vlan 100                     //端口E1/3属于VLAN100
Interface Ethernet1/4
 switchport access vlan 100
略
Interface Ethernet1/11
 switchport access vlan 100
Interface Ethernet1/12
 switchport access vlan 100                     //端口E1/12属于VLAN100
Interface Ethernet1/13
 switchport access vlan 200                     //端口E1/13属于VLAN200
Interface Ethernet1/14
 switchport access vlan 200                     //端口E1/14属于VLAN200
Interface Ethernet1/15
 switchport access vlan 200                     //端口E1/15属于VLAN200
略
Interface Ethernet1/23
```

```
   switchport access vlan 200              //端口E1/23属于VLAN200
  Interface Ethernet1/24
   switchport access vlan 200              //端口E1/24属于VLAN200
  Interface Ethernet1/25
  Interface Ethernet1/26
  Interface Ethernet1/27
  Interface Ethernet1/28
  interface Vlan1
   ip address 192.168.1.1 255.255.255.0
  no login
  End
```

2.3 实训三 跨交换机相同 VLAN 访问

任务描述

某学校教学楼有两层，分别是一年级、二年级，每个楼层都有一台交换机满足老师的上网需求；每个年级都有电子商务部和网络技术部两个部的学生。现在要求两个年级的电子商务部的计算机可以互相访问，两个年级的网络技术部的计算机可以互相访问，电子商务部和网络技术部不可以自由访问。分别设置交换机A和交换机B，1~8端口属于VLAN 100，9~16端口属于VLAN 200，24端口为Trunk端口。

本实训需要设备如下：

DCS-3950 交换机两台，PC 四台，Console 线缆一根，直通网线五根。PC 的 IP 地址、PC 所连交换机端口和所属 VLAN 如表2-2 所示。

表2-2 PC的IP地址、所连交换机端口和所属VLAN

设备名称	IP 地址	所连交换机	所属 VLAN	连接端口
PC1	192.168.1.101/24	SwitchA	VLAN100	E1/1
PC2	192.168.1.201/24	SwitchA	VLAN200	E1/9
PC3	192.168.1.102/24	SwitchB	VLAN100	E1/1
PC4	192.168.1.202/24	SwitchB	VLAN200	E1/9

网络拓扑图如图 2-3 所示。

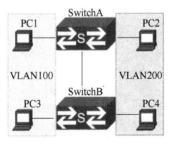

图 2-3 网络拓扑图

任务准备

交换机上的 VLAN 只具有本地意义，要想在不同交换机间同一 VLAN 计算机通信，就用到了 IEEE 802.1Q 协议，它在以太网帧的基础上增加了标签(tag)，其中包含 VLAN ID 号。

Untag 报文就是普通的 ethernet 报文，普通 PC 的网卡可以识别这样的报文进行通信。

Tag 报文结构的变化是在源 mac 地址和目的 mac 地址之后，加上了 4Bytes 的 VLAN 信息，也就是 VLAN Tag 头，这样的报文是普通 PC 的网卡不能识别的，交换机能识别并进行处理。

神州数码交换机的以太网端口有两种链路类型：Access 和 Trunk。

Access 类型的端口只能属于 1 个 VLAN，一般用于连接计算机的端口。

Trunk 类型的端口可以允许多个 VLAN 通过，可以接收和发送多个 VLAN 的报文，一般用于交换机之间连接的端口。

默认 VLAN：Access 端口只属于 1 个 VLAN，所以它的默认 VLAN 就是它所在的 VLAN，不用设置；Trunk 端口属于多个 VLAN，所以需要设置默认 VLAN ID。默认情况下，Trunk 端口的默认 VLAN 为 VLAN1。如果设置了端口的默认 VLAN ID，当端口接收到不带 VLAN Tag 的报文后，则将报文转发到属于默认 VLAN 的端口；当端口发送带有 VLAN Tag 的报文时，如果该报文的 VLAN ID 与端口默认的 VLAN ID 相同，则系统将去掉报文的 VLAN Tag，然后再发送该报文。

Access 端口接收报文操作：收到一个报文，判断是否有 VLAN 信息。如果没有，则打上端口的 PVID，并进行交换转发；如果有，则直接丢弃。

Access 端口发送报文操作：将报文的 VLAN 信息剥离，直接发送出去。

Trunk 端口收报文操作：收到一个报文，判断是否有 VLAN 信息。如果有，判断该 Trunk 端口是否允许该 VLAN 的数据进入，如果可以则转发，否则丢弃；如果没有 VLAN 信息则打上端口的 PVID，并进行交换转发。

Trunk 端口发报文操作：比较将要发送报文的 VLAN 信息和端口的 PVID，如果不相等则直接发送，如果两者相等则剥离 VLAN 信息，再发送。

任务实施

SwitchA

1. 给交换机设置标识符。

```
DCS-3950-28C>enable
DCS-3950-28C#config
DCS-3950-28C(config)#hostname SwitchA          //设置交换机标识符为 SwitchA
```

2. 在交换机中添加 VLAN100 和 VLAN200，并添加端口。

```
SwitchA(config)#vlan 100                       //创建 VLAN100
SwitchA(config-vlan100)#switchport interface ethernet 1/1~8
                                               //端口 1~8 加入 VLAN100
Set the port Ethernet1/1 access vlan 100 successfully
Set the port Ethernet1/2 access vlan 100 successfully
```

```
Set the port Ethernet1/3 access vlan 100 successfully
Set the port Ethernet1/4 access vlan 100 successfully
Set the port Ethernet1/5 access vlan 100 successfully
Set the port Ethernet1/6 access vlan 100 successfully
Set the port Ethernet1/7 access vlan 100 successfully
Set the port Ethernet1/8 access vlan 100 successfully
SwitchA(config)#vlan 200                           //创建VLAN100
SwitchA(config-vlan200)#switchport interface ethernet 1/9~16
                                                   //端口9~16加入VLAN100
Set the port Ethernet1/9 access vlan 200 successfully
Set the port Ethernet1/10 access vlan 200 successfully
Set the port Ethernet1/11 access vlan 200 successfully
Set the port Ethernet1/12 access vlan 200 successfully
Set the port Ethernet1/13 access vlan 200 successfully
Set the port Ethernet1/14 access vlan 200 successfully
Set the port Ethernet1/15 access vlan 200 successfully
Set the port Ethernet1/16 access vlan 200 successfully
```

3．设置24端口为Trunk端口。

```
SwitchA(config)#interface e1/24
SwitchA(config-if-ethernet1/24)#switchport mode trunk   //设置端口24为
Trunk端口
Set the port Ethernet1/24 mode Trunk successfully
SwitchA(config-if-ethernet1/24)#switchport trunk allowed vlan all
                                   //此Trunk端口允许所有VLAN通过
```

4．查看交换机配置。

```
SwitchA(config)#show running-config
no service password-encryption
hostname SwitchA
sysLocation China
sysContact 800-810-9119
username admin privilege 15 password 0 admin
vlan 1;100;200                      //交换机现有VLAN1、VLAN100和VLAN200
Interface Ethernet1/1
 switchport access vlan 100         //端口1属于VLAN100
 Interface Ethernet1/2
 switchport access vlan 100
 Interface Ethernet1/3
 switchport access vlan 100
 Interface Ethernet1/4
 switchport access vlan 100
 Interface Ethernet1/5
 switchport access vlan 100
 Interface Ethernet1/6
 switchport access vlan 100
```

```
Interface Ethernet1/7
 switchport access vlan 100
Interface Ethernet1/8
 switchport access vlan 100
Interface Ethernet1/9
 switchport access vlan 200                    //端口 9 属于 VLAN200
Interface Ethernet1/10
 switchport access vlan 200
Interface Ethernet1/11
 switchport access vlan 200
Interface Ethernet1/12
 switchport access vlan 200
Interface Ethernet1/13
 switchport access vlan 200
Interface Ethernet1/14
 switchport access vlan 200
Interface Ethernet1/15
 switchport access vlan 200
Interface Ethernet1/16
 switchport access vlan 200
Interface Ethernet1/17
Interface Ethernet1/18
Interface Ethernet1/19
Interface Ethernet1/20
Interface Ethernet1/21
Interface Ethernet1/22
Interface Ethernet1/23
Interface Ethernet1/24
 switchport mode trunk      //此端口为 Trunk 端口，默认允许所有 VLAN 数据通过
Interface Ethernet1/25
Interface Ethernet1/26
Interface Ethernet1/27
Interface Ethernet1/28
interface Vlan1
no login
End
```

SwitchB

1. 给交换机设置标识符。

```
DCS-3950-28C>enable
DCS-3950-28C#config
DCS-3950-28C(config)#hostname SwitchB        //设置交换机标识符为 SwitchB
```

2. 在交换机中添加 VLAN100 和 VLAN200，并添加端口。

```
SwitchB(config)#vlan 100                                   //创建 VLAN100
SwitchB(config-vlan100)#switchport interface ethernet 1/1-8
```

```
            //端口1~8加入VLAN100
Set the port Ethernet1/1 access vlan 100 successfully
Set the port Ethernet1/2 access vlan 100 successfully
Set the port Ethernet1/3 access vlan 100 successfully
Set the port Ethernet1/4 access vlan 100 successfully
Set the port Ethernet1/5 access vlan 100 successfully
Set the port Ethernet1/6 access vlan 100 successfully
Set the port Ethernet1/7 access vlan 100 successfully
Set the port Ethernet1/8 access vlan 100 successfully
SwitchB(config)#vlan 200                              //创建VLAN200
SwitchB(config-vlan200)#switchport interface ethernet 1/9-16
            //端口9~16加入VLAN100
Set the port Ethernet1/9 access vlan 200 successfully
Set the port Ethernet1/10 access vlan 200 successfully
Set the port Ethernet1/11 access vlan 200 successfully
Set the port Ethernet1/12 access vlan 200 successfully
Set the port Ethernet1/13 access vlan 200 successfully
Set the port Ethernet1/14 access vlan 200 successfully
Set the port Ethernet1/15 access vlan 200 successfully
Set the port Ethernet1/16 access vlan 200 successfully
```

3．设置24端口为Trunk端口。

```
SwitchB(config)#interface e1/24
SwitchB(config-if-ethernet1/24)#switchport mode trunk
Set the port Ethernet1/24 mode Trunk successfully
SwitchB(config-if-ethernet1/24)#switchport trunk allowed vlan all
```

4．查看交换机配置。

```
SwitchB(config)#show running-config
no service password-encryption
hostname SwitchB
sysLocation China
sysContact 800-810-9119
username admin privilege 15 password 0 admin
vlan 1;100;200
Interface Ethernet1/1
 switchport access vlan 100
Interface Ethernet1/2
 switchport access vlan 100
Interface Ethernet1/3
 switchport access vlan 100
Interface Ethernet1/4
 switchport access vlan 100
Interface Ethernet1/5
 switchport access vlan 100
Interface Ethernet1/6
```

```
  switchport access vlan 100
Interface Ethernet1/7
  switchport access vlan 100
Interface Ethernet1/8
  switchport access vlan 100
Interface Ethernet1/9
  switchport access vlan 200
Interface Ethernet1/10
  switchport access vlan 200
Interface Ethernet1/11
  switchport access vlan 200
Interface Ethernet1/12
  switchport access vlan 200
Interface Ethernet1/13
  switchport access vlan 200
Interface Ethernet1/14
  switchport access vlan 200
Interface Ethernet1/15
  switchport access vlan 200
Interface Ethernet1/16
  switchport access vlan 200
Interface Ethernet1/17
Interface Ethernet1/18
Interface Ethernet1/19
Interface Ethernet1/20
Interface Ethernet1/21
Interface Ethernet1/22
Interface Ethernet1/23
Interface Ethernet1/24
  switchport mode trunk
Interface Ethernet1/25
Interface Ethernet1/26
Interface Ethernet1/27
Interface Ethernet1/28
interface Vlan1
no login
```

这两台交换机配置完成后，同属于 VLAN100 的 PC1 和 PC3 之间能互相 ping 通，同属于 VLAN200 的 PC2 和 PC4 之间能互相 ping 通，证明了跨交换机的相同 VLAN 内 PC 能通信。具体 ping 的结果，由于篇幅所限，在此就不再赘述。

2.4 实训四　交换机私有 VLAN 的应用

任务描述

某小区的宽带接入商近日不断接到一些居民小区用户投诉网速慢，经过检测是由于本小区的一些 PC 运行网络游戏占用大量带宽导致其他机器网速下降。为了解决这一问题，而又不影响现有局域网对战游戏的用户，接入商决定启用交换机的私有 VLAN，建立 VLAN10、VLAN20 和 VLAN30。VLAN10 的 IP 地址为"192.168.10.1/24"，VLAN 类型为 primary，端口为 1~5 端口，此 VLAN 是连接服务群和连接 Internet。VLAN20 的 VLAN 类型为 community，端口为 6~10 端口，此 VLAN 是为有局域网游戏对战要求的用户提供服务。VLAN30 的 VLAN 类型为 isolated，端口为 11~15 端口，此 VLAN 为有更高安全要求的用户提供服务。

网络拓扑图如图 2-4 所示。

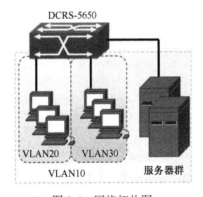

图 2-4　网络拓扑图

任务准备

PVLAN 技术：PVLAN（Private VLAN）是一种新的 VLAN 机制，所有服务器在同一个子网中，但服务器只能与自己的默认网关通信。PVLAN 的应用对于保证接入网络的数据通信的安全性是非常有效的，用户只需与自己的默认网关连接，一个 PVLAN 不需要多个 VLAN 和 IP 子网就提供了具备第 2 层数据通信安全性的连接，所有的用户都接入 PVLAN，从而实现了所有用户与默认网关的连接，而与 PVLAN 内的其他用户没有任何访问。PVLAN 功能可以保证同一个 VLAN 中的各个端口相互之间不能通信。这样即使同一 VLAN 中的用户，相互之间也不会受到广播的影响。PVLAN 采用两层 VLAN 隔离技术，只有上层 VLAN 全局可见，下层 VLAN 相互隔离。PVLAN 通常用于企业内部网，用来防止连接到某些接口或接口组的网络设备之间的相互通信，但却允许与默认网关进行通信。尽管各设备处于不同的 PVLAN 中，它们可以使用相同的 IP 子网。

每个 PVLAN 包含 2 种 VLAN：主 VLAN（Primary VLAN）和辅助 VLAN（Secondary

VLAN)。辅助 VLAN 包含两种类型：隔离 VLAN（isolated VLAN）和团体 VLAN（community VLAN）。

PVLAN 通信范围分为两类。Primary VLAN 可以和所有它所关联的 isolated VLAN，community VLAN 通信。Community VLAN 可以同那些处于相同 community VLAN 内的 community 端口通信，也可以与 PVLAN 中的 promiscuous 端口通信。每个 PVLAN 可以有多个 community VLAN。Isolated VLAN 不可以和处于相同 isolated VLAN 内的其他 isolated 端口通信，只可以与 promiscuous 端口通信。每个 PVLAN 中只能有一个 isolated VLAN。

任务实施

1. 创建相应 VLAN，设置不同 VLAN 类型，做主从 VLAN 之间的关联。

```
DCS-3950-28C>enable
DCS-3950-28C#config
DCS-3950-28C(config)#vlan 10                         //创建 VLAN10
DCS-3950-28C(config-vlan10)#private-vlan primary     //设置为主 VLAN
DCS-3950-28C(config-vlan10) #exit
DCS-3950-28C(config)#vlan 20                         //创建 VLAN20
DCS-3950-28C(config-vlan20)#private-vlan community   //设置为团体 VLAN
DCS-3950-28C(config-vlan20) #exit
DCS-3950-28C(config)#vlan 30                         //创建 VLAN30
DCS-3950-28C(config-vlan30)#private-vlan isolated    //设置为隔离 VLAN
DCS-3950-28C(config-vlan30) #exit
DCS-3950-28C(config)#vlan 10
DCS-3950-28C(config-vlan10)#private-vlan association 20;30
                          //主 VLAN10 关联到团体 VLAN20 和隔离 VLAN30
Set vlan 10 associated vlan successfully
```

2. 添加相应端口。

```
DCS-3950-28C(config)vlan 10
DCS-3950-28C(config-vlan10)#switchport interface e1/1-5    //端口 1~5 加入
VLAN10
DCS-3950-28C(config-vlan10) #exit
DCS-3950-28C(config)vlan 20
DCS-3950-28C(config-vlan20)#switchport interface e1/6-10   //端口 6~10 加
入 VLAN20
DCS-3950-28C(config-vlan20)#exit
DCS-3950-28C(config)vlan 30
DCS-3950-28C(config-vlan30)#switchport interface e1/11-15  //端口 11~15
加入 VLAN30
DCS-3950-28C(config-vlan30)#exit
```

3. 查看交换机配置。

```
DCS-3950-28C(config)#show running-config
no service password-encryption
hostname DCS-3950-28C
sysLocation China
```

```
 sysContact 800-810-9119
 username admin privilege 15 password 0 admin
 vlan 1
 vlan 20
  private-vlan community                         //VLAN20 为团体 VLAN
 vlan 30
  private-vlan isolated                          //VLAN30 为隔离 VLAN
 vlan 10
  private-vlan primary                           //VLAN10 为主 VLAN
private-vlan association 20;30                   //关联到辅助 VLAN20 和 VLAN30
   Interface Ethernet1/1
  Interface Ethernet1/2
  Interface Ethernet1/3
  Interface Ethernet1/4
  Interface Ethernet1/5
  Interface Ethernet1/6
  Interface Ethernet1/7
  Interface Ethernet1/8
  Interface Ethernet1/9
  Interface Ethernet1/10
  Interface Ethernet1/11
  Interface Ethernet1/12
  Interface Ethernet1/13
  Interface Ethernet1/14
  Interface Ethernet1/15
  Interface Ethernet1/16
  Interface Ethernet1/17
  Interface Ethernet1/18
  Interface Ethernet1/19
  Interface Ethernet1/20
  Interface Ethernet1/21
  Interface Ethernet1/22
  Interface Ethernet1/23
  Interface Ethernet1/24
  Interface Ethernet1/25
  Interface Ethernet1/26
  Interface Ethernet1/27
  Interface Ethernet1/28
  interface Vlan1
   ip address 192.168.1.1 255.255.255.0
  no login
  End
```

4．验证配置。

```
DCS-3950-28C(config)# show vlan private-vlan
VLAN Name         Type        Asso VLAN Ports
```

```
10      VLAN0010      Primary      20   30
20      VLAN0020      Community    10
30      VLAN0030      Isolate      10
```

本交换机配置完成后,可以通过 4 中的显示效果看到,VLAN10 关联到辅助 VLAN20 和 VLAN30,同时通过不同 PC 间互 ping,VLAN20 和 VLAN30 中 PC 都能 ping 通 VLAN10 内服务器,VLAN20 和 VLAN30 中 PC 互相不能 ping 通,VLAN20 内 PC 间能 ping 通,VLAN30 内 PC 不能 ping 通。

2.5　实训五　提升交换机间的连接带宽

任务描述

某公司有两台交换机通过端口 1 相连,此链路带宽为 100Mb/s。管理员发现此链路成为了网络性能的瓶颈,为解决此问题,管理员把 DCS-A 的 "E1/1-2" 与 DCS-B 的 "E1/1-2" 进行端口汇聚,形成手工生成链路汇聚组 1,增加了交换机间链路带宽。

网络拓扑图如图 2-5 所示。

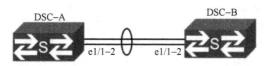

图 2-5　网络拓扑图

任务准备

神州数码交换机提供了两种配置端口汇聚的方法:手工生成 Port Channel,LACP(Link Aggregation Control Protocol)动态生成 Port Channel。只有双工模式为全双工模式的端口才能进行端口汇聚。

为使 Port Channel 正常工作,交换机 Port Channel 的成员端口必须具备以下相同的属性:端口均为全双工模式;端口速率相同;端口同为 Access 端口并且属于同一个 VLAN 或同为 Trunk 端口或同为 hybrid 端口;如果端口同为 Trunk 端口或同为 hybrid 端口,则其 Allowed VLAN 和 Native VLAN 属性也应该相同。

当交换机以手工方式配置 Port Channel 或以 LACP 方式动态生成 Port Channel,系统将自动选举出 Port Channel 中端口号最小的端口作为 Port Channel 的主端口(Master Port)。若交换机打开 Spanning-tree 功能,Spanning-tree 将把 Port Channel 视为一个逻辑端口,并且由主端口发送 BPDU 帧。

另外,端口汇聚功能的实现与交换机所使用的硬件有密切关系,本实训中使用的交换机支持任意两个相同属性的物理端口的汇聚,最大组数为 15 个,组内最多的端口数为 8 个。

汇聚端口一旦汇聚成功就可以把它当成一个普通的端口使用,在交换机中还建立了汇聚接口配置模式,与 VLAN 和物理接口配置模式一样,用户能在汇聚接口配置模式下对汇聚端口进行相关的配置。

任务实施

DCS-A

1. 创建链路汇聚组。

```
DCS-3950-28C>enable
DCS-3950-28C#config
DCS-3950-28C(config)#hostname DCS-A
DCS-A(config)#port-group 1                              //创建链路汇聚组1
```

2. 将端口"E1/1-2"加入链路汇聚组。

```
DCS-A(config)#interface e1/1-2
DCS-A(config-if-port-range)#port-group 1 mode on
                    //把1-2端口加入汇聚组1,并且汇聚类型为手工汇聚
DCS-A(config-if-port-range)#exit
```

3. 查看交换机配置。

```
DCS-A(config)#show running-config
no service password-encryption
hostname DSC-A
sysLocation China
sysContact 800-810-9119
username admin privilege 15 password 0 admin
vlan 1
port-group 1
Interface Ethernet1/1
 port-group 1 mode on                    //端口属于汇聚组1,汇聚类型为手工汇聚
Interface Ethernet1/2
 port-group 1 mode on                    //端口属于汇聚组1,汇聚类型为手工汇聚
Interface Ethernet1/3
Interface Ethernet1/4
略
Interface Ethernet1/25
Interface Ethernet1/26
Interface Ethernet1/27
Interface Ethernet1/28
Interface Port-Channel1                   //逻辑端口汇聚组1
interface Vlan1
no login
End
```

DCS-B

1. 创建链路汇聚组。

```
DCS-3950-28C>enable
```

```
DCS-3950-28C#config
DCS-3950-28C(config)#hostname DCS-B
DCS-B(config)#port-group 1                                    //创建链路汇聚组1
```

2. 将端口"E1/1-2"加入链路汇聚组。

```
DCS-B(config)#interface e1/1-2
DCS-B(config-if-port-range)#port-group 1 mode on
                    //把1-2端口加入汇聚组1,并且汇聚类型为手工汇聚
DCS-B(config-if-port-range)#exit
```

3. 查看交换机配置。

```
DCS-B(config)#show running-config
no service password-encryption
hostname DSC-B
sysLocation China
sysContact 800-810-9119
username admin privilege 15 password 0 admin
vlan 1
port-group 1
Interface Ethernet1/1
 port-group 1 mode on          //端口属于汇聚组1,汇聚类型为手工汇聚
Interface Ethernet1/2
 port-group 1 mode on          //端口属于汇聚组1,汇聚类型为手工汇聚
Interface Ethernet1/3
Interface Ethernet1/4
略
Interface Ethernet1/25
Interface Ethernet1/26
Interface Ethernet1/27
Interface Ethernet1/28
Interface Port-Channel1         //逻辑端口汇聚组1
interface Vlan1
no login
End
```

4. 显示汇聚组状态。

```
DCS-3950-28C(config)#show port-group 1 detail
Flags: A -- LACP_Activity, B -- LACP_timeout, C -- Aggregation,
       D -- Synchronization, E -- Collecting, F -- Distributing,
       G -- Defaulted, H -- Expired
Port-group number: 1,  Mode: on,   Load-balance: src-mac
```

2.6 实训六 交换机端口镜像

任务描述

某公司有一台服务器连接在交换机的端口1。管理员最近发现Web服务器经常遭到攻击，为了对服务器上发送及接收的数据流进行监控，而又不影响服务器正常工作，管理员通过设置端口镜像，在端口2复制端口1的发送和接收的数据，并在端口2连接的PC2上安装sniffer来分析数据。

网络拓扑图如图2-6所示。

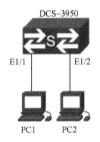

图2-6 网络拓扑图

任务准备

端口镜像：端口镜像是把交换机一个或多个端口（VLAN）的数据镜像到一个或多个端口的方法。本实训采用的交换机可以建立多个镜像组，每个镜像组中镜像的源端口只能有一个，而镜像组中的目的端口可以有多个。

任务实施

1. 设置镜像源端口。

```
DCS-3950-28C>enable
DCS-3950-28C#config
DCS-3950-28C(config)#monitor session 1 source interface e1/1 both
          //设置E1/1为镜像源端口,并且监控接口的进出方向数据
```

2. 设置镜像目的端口。

```
DCS-3950-28C(config)#monitor session 1 destination interface e1/2
          //设置E1/2为镜像目的端口
```

3. 查看交换机配置。

```
DCS-3950-28C(config)#show running-config
no service password-encryption
hostname DCS-3950-28C
sysLocation China
sysContact 800-810-9119
username admin privilege 15 password 0 admin
vlan 1
```

```
Interface Ethernet1/1
Interface Ethernet1/2
略
Interface Ethernet1/25
Interface Ethernet1/26
Interface Ethernet1/27
Interface Ethernet1/28
interface Vlan1
no login
monitor session 1 source interface Ethernet1/1 rx
                //设置E1/1为镜像源端口,并且监控端口的进方向数据
monitor session 1 source interface Ethernet1/1 tx
                //设置E1/1为镜像源端口,并且监控端口的出方向数据
monitor session 1 destination interface Ethernet1/2
                //设置E1/2为镜像目的端口
end
```

经过以上步骤后,端口2所连接的PC2就可以通过sniffer软件来监控服务器PC1上的数据流。对PC2中sniffer部分的内容在此不再赘述。

2.7 实训七 避免网络的冗余链路危害

任务描述

某公司网络中有三台交换机,为了实现链路冗余,三台交换机间都有链路相连,但此时出现了广播风暴的问题。管理员决定启用生成树协议来解决广播风暴,同时又保持链路间冗余,当某条链路失败后,其他链路能自动启用。

公司交换机间端口连接表见表2-3。

表2-3 交换机间端口连接表

端口号	所连端口
SwitchA E1/1	SwitchB E1/2
SwitchA E1/2	SwitchC E1/1
SwitchB E1/1	SwitchC E1/2

网络拓扑图如图2-7所示。

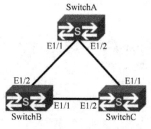

图2-7 网络拓扑图

任务准备

STP：STP（Spanning Tree Protocol）即生成树算法的网桥协议，定义在 IEEE 802.1D 中，是一种链路管理协议。生成树协议的主要功能有两个：一是在以太网络中利用生成树算法，创建一个以某台交换机为根的生成树，避免环路。二是当以太网络拓扑发生变化时，通过生成树协议达到收敛保护的目的。

802.1D 生成树协议概述：802.1D 生成树协议包括网桥标识（bridge ID），即网桥优先级（4bits）+ 系统标识（VLAN ID；12bits）+ MAC 地址。下面简单进行介绍。

网桥协议数据单元（BPDU）：在竞选根网桥阶段每个网桥都会发送，假设自己就是根网桥。竞选成功收敛后，BPDU 只从根网桥发出，其他网桥在根端口接收后向下中继。

拓扑改变提示（TCN）BGDU：当拓扑发生变化时，其他网桥可以从根端口发出该 BPDU，到达根网桥。根网桥在配置 BPDU 中设定 TCN 位，提示其他网桥快速清理 MAC 地址表。

时间值：HELLO 间隔为 2s，BPDU 发送间隔。MAX AGE 间隔为 20s，BPDU 的保留时间。FWD_DELAY 间隔为 15s，监听（listening）和学习（learning）的时间。

路径代价：与链路速率相关，用于计算网桥间的距离。

端口状态有以下 5 种状态。①关闭（disable）：端口处于管理关闭状态。②阻塞（blocking）：不能转发用户数据。③监听（listening）：接口开始启动。④学习（learning）：学习 MAC 地址，构建 MAC 表进程项。⑤转发（forwarding）：可以转发用户数据。

根网桥选择标准：最低的网桥标识号。

任务实施

SwitchA

1. 创建交换机名。

```
DCS-3950-28C>enable
DCS-3950-28C#config
DCS-3950-28C(config)#hostname SwitchA
```

2. 启用生成树，并配置。

```
SwitchA(config)#spanning-tree                          //启动生成树协议
MSTP is starting now, please wait..........
MSTP is enabled successfully.
SwitchA(config)#spanning-tree mode stp                 //设置生成树协议模式为STP
SwitchA(config)#spanning-tree priority 4096            //设置本交换机的STP优先级为4096
```

3. 查看交换机配置。

```
SwitchA(config)#show running-config
no service password-encryption
hostname SwitchA
sysLocation China
sysContact 800-810-9119
username admin privilege 15 password 0 admin
```

```
spanning-tree                              //启动生成树协议
spanning-tree mode stp                     //生成树协议模式为STP
spanning-tree priority 4096                //本交换机的STP优先级为4096
vlan 1
Interface Ethernet1/1
Interface Ethernet1/2
Interface Ethernet1/3
Interface Ethernet1/4
Interface Ethernet1/5
Interface Ethernet1/6
Interface Ethernet1/7
Interface Ethernet1/8
Interface Ethernet1/9
Interface Ethernet1/10
Interface Ethernet1/11
Interface Ethernet1/12
Interface Ethernet1/13
Interface Ethernet1/14
Interface Ethernet1/15
Interface Ethernet1/16
Interface Ethernet1/17
Interface Ethernet1/18
Interface Ethernet1/19
Interface Ethernet1/20
Interface Ethernet1/21
Interface Ethernet1/22
Interface Ethernet1/23
Interface Ethernet1/24
Interface Ethernet1/25
Interface Ethernet1/26
Interface Ethernet1/27
Interface Ethernet1/28
no login
End
```

SwitchB

1. 创建交换机名。

```
DCS-3950-28C>enable
DCS-3950-28C#config
DCS-3950-28C(config)#hostname SwitchB
```

2. 启用生成树，并配置。

```
SwitchB(config)#spanning-tree                //启动生成树协议
MSTP is starting now, please wait..........
```

```
MSTP is enabled successfully.
SwitchB(config)#spanning-tree mode stp        //设置生成树协议模式为STP
SwitchB(config)#spanning-tree priority 8192   //设置本交换机的STP优先级为8192
```

3. 查看交换机配置。

```
SwitchB(config)#show running-config
no service password-encryption
hostname SwitchB
sysLocation China
sysContact 800-810-9119
username admin privilege 15 password 0 admin
spanning-tree                                 //启动生成树协议
spanning-tree mode stp                        //生成树协议模式为STP
spanning-tree priority 8192                   //本交换机的STP优先级为8192
vlan 1
Interface Ethernet1/1
Interface Ethernet1/2
Interface Ethernet1/3
Interface Ethernet1/4
Interface Ethernet1/5
Interface Ethernet1/6
Interface Ethernet1/7
Interface Ethernet1/8
Interface Ethernet1/9
Interface Ethernet1/10
Interface Ethernet1/11
Interface Ethernet1/12
Interface Ethernet1/13
Interface Ethernet1/14
Interface Ethernet1/15
Interface Ethernet1/16
Interface Ethernet1/17
Interface Ethernet1/18
Interface Ethernet1/19
Interface Ethernet1/20
Interface Ethernet1/21
Interface Ethernet1/22
Interface Ethernet1/23
Interface Ethernet1/24
Interface Ethernet1/25
Interface Ethernet1/26
Interface Ethernet1/27
Interface Ethernet1/28
no login
End
```

SwitchC

1. 创建交换机名称。

```
DCS-3950-28C>enable
DCS-3950-28C#config
DCS-3950-28C(config)#hostname SwitchC
```

2. 启用生成树,并配置。

```
SwitchC(config)#spanning-tree                          //启动生成树协议
MSTP is starting now, please wait...........
MSTP is enabled successfully.
SwitchC(config)#spanning-tree mode stp                 //生成树协议模式为STP
SwitchC(config)#spanning-tree priority 8192            //本交换机的STP优先级为8192
```

3. 查看交换机配置。

```
SwitchC(config)#show running-config
no service password-encryption
hostname SwitchC
sysLocation China
sysContact 800-810-9119
username admin privilege 15 password 0 admin
spanning-tree                                          //启动生成树协议
spanning-tree mode stp                                 //生成树协议模式为STP
spanning-tree priority 8192                            //本交换机的STP优先级为8192
vlan 1
Interface Ethernet1/1
Interface Ethernet1/2
Interface Ethernet1/3
Interface Ethernet1/4
Interface Ethernet1/5
Interface Ethernet1/6
Interface Ethernet1/7
Interface Ethernet1/8
Interface Ethernet1/9
Interface Ethernet1/10
Interface Ethernet1/11
Interface Ethernet1/12
Interface Ethernet1/13
Interface Ethernet1/14
Interface Ethernet1/15
Interface Ethernet1/16
Interface Ethernet1/17
Interface Ethernet1/18
Interface Ethernet1/19
Interface Ethernet1/20
Interface Ethernet1/21
Interface Ethernet1/22
```

```
Interface Ethernet1/23
Interface Ethernet1/24
Interface Ethernet1/25
Interface Ethernet1/26
Interface Ethernet1/27
Interface Ethernet1/28
interface Vlan1
 ip address 192.168.1.3 255.255.255.0
no login
End
```

所有配置完成后查看交换机中生成树表项。

```
SwitchA
SwitchA(config)#show spanning-tree
          -- STP Bridge Config Info --
Standard      :  IEEE 802.1d
Bridge MAC    :  00:03:0f:29:1d:71
Bridge Times  :  Max Age 20, Hello Time 2, Forward Delay 15
Force Version:  0
###############################################################
Self Bridge Id   : 4096.00:03:0f:29:1d:71       //本交换机的优先级为4096
Root Id          : this switch                  //本交换机为根交换机
Ext.RootPathCost : 0
Root Port ID     : 0
   PortName       ID      ExtRPC State Role   DsgBridge         DsgPort
   ---------------------------------------------------------------------
   Ethernet1/1  128.001      0   FWD   DSGN   4096.00030f291d71  128.001
   Ethernet1/2  128.002      0   FWD   DSGN   4096.00030f291d71  128.002
```

可以看到本交换机中为生成树的根，同时 Ethernet1/1 和 Ethernet1/2 的状态为 FWD，为指定端口。

```
SwitchB(config)#show spanning-tree
          -- STP Bridge Config Info --
Standard      :  IEEE 802.1d
Bridge MAC    :  00:03:0f:29:1d:65
Bridge Times  :  Max Age 20, Hello Time 2, Forward Delay 15
Force Version:  0
###############################################################
Self Bridge Id   : 8192.00:03:0f:29:1d:65       //本交换机的优先级为8192
Root Id          : 4096.00:03:0f:29:1d:71       //根交换机为SwitchA
Ext.RootPathCost : 199999
Root Port ID     : 128.2
   PortName       ID      ExtRPC State Role   DsgBridge         DsgPort
   ---------------------------------------------------------------------
   Ethernet1/1  128.001   199999  FWD   DSGN   8192.00030f291d65  128.001
   Ethernet1/2  128.002        0  FWD   ROOT   4096.00030f291d71  128.001
```

可以看到 SwitchA 为生成树的根，同时 Ethernet1/1 的状态为 FWD（转发），为指定端口，是所属网段中离根最近的端口。SwitchB 中 Ethernet1/2 的状态为 FWD，为根端口，为本交换机离根网桥最近的端口。

```
SwitchC(config)#show spanning-tree
             -- STP Bridge Config Info --
Standard       : IEEE 802.1s
Bridge MAC     : 00:03:0f:14:bd:59
Bridge Times   : Max Age 20, Hello Time 2, Forward Delay 15
Force Version: 0
###############################################################
Self Bridge Id    : 8192.00:03:0f:14:bd:59      //本交换机的优先级为8192
Root Id           : 4096.00:03:0f:29:1d:71      //根交换机为 SwitchA
Ext.RootPathCost  : 199999
Root Port ID      : 128.1
   PortName       ID      ExtRPC  State Role    DsgBridge        DsgPort
   ------------  -------  ------  ----- ----    ----------------  -------
   Ethernet1/1   128.001       0  FWD   ROOT    4096.00030f291d71 128.002
   Ethernet1/2   128.002  199999  BLK   ALTR    8192.00030f291d65 128.001
```

可以看到 SwitchA 为生成树的根，同时 SwitchC 的 Ethernet1/1 的状态为 FWD（转发），为根端口，是本交换机离根网桥最近的端口。SwitchC 的 Ethernet1/2 的状态为 BLK（阻塞状态），为 ALTR 端口，不是所属网段中离根最近的端口。

项目 3

三层交换机的应用

教学目标

通过本章的学习,学生可以了解三层交换机的基础配置命令以及针对基础配置的方式和方法。掌握三层交换机针对VLAN间路由的设置以及针对三层交换机的路由功能的设置。

能力目标

了解三层交换机的作用
熟悉VLAN间路由的配置
掌握三层交换机的路由功能的配置

知识目标

熟悉交换机针对子网的设置
熟悉交换机针对子网间通信的配置
熟悉三层交换机各种路由功能的配置

主要教学内容

VLAN间路由的设置
各种路由协议在三层交换机中的设置
三层交换机特殊功能的设置

3.1 实训一 三层交换机 VLAN 的划分与 VLAN 间路由

任务描述

某中学同一楼层有两个一体化教室,这两个教室的信息端口都连接在一台三层交换机上。为了保证两个一体化教室的相对独立,学校为这两个一体化教室分配了不同的 IP 地址段,要求它们之间既不在同一广播域,又能相互通信。管理员划分了 VLAN1 和 VLAN2,端口 1~10 属于 VLAN1,端口 11~20 属于 VLAN2,VLAN1 的 IP 地址为 "192.168.1.1/24",VLAN2 的 IP 地址为 "192.168.2.1/24"。

本任务采用如下设备:DCS-5650 交换机一台,PC 两台,Console 线缆一根,直通网线两根。

网络拓扑图如图 3-1 所示。

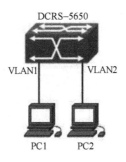

图 3-1 网络拓扑图

任务准备

路由是指路由器从一个端口上收到数据包,根据数据包的目的地址进行定向并转发到另一个端口的过程。

三层交换机是具有部分路由器功能的交换机,路由功能能够做到一次路由、多次转发,加快大型局域网内部的数据交换。对于数据包转发等规律性的过程由硬件高速实现,而像路由信息更新、路由表维护、路由计算、路由确定等功能,由软件实现。三层交换技术就是二层交换技术+三层转发技术。传统交换技术是数据链路层进行操作的,而三层交换技术是在网络模型中的第三层实现了数据包的高速转发,既可实现网络路由功能,又可根据不同网络状况做到最优网络性能。

在三层交换机上可以创建三层端口。三层端口并不是实际的物理接口,是在 VLAN 的基础上创建的,它是一个虚拟的端口。三层端口可以包含一个或多个二层端口,但也可以不包含任何二层端口。三层端口包含的二层端口中,需要至少有一个是 UP 状态,三层端口才是 UP 状态,否则为 DOWN 状态。交换机中所有的三层端口默认使用一个相同的 MAC 地址,此地址是在三层端口创建时从交换机保留的 MAC 地址中选取的。在三层端口上可以配置 IP 地址,交换机可以通过配置在三层端口上的 IP 地址,与其他设备进行 IP 协议的传输。交换机也可以在不同的三层端口之间转发 IP 协议报文。

任务实施

1. 创建 VLAN1 和 VLAN2，并加入相应端口。

```
DCRS-5650-28>enable
DCRS-5650-28#config
DCRS-5650-28(config)#vlan 1
DCRS-5650-28(config-Vlan1)#switchport interface e0/0/1-10
                                         //端口1～10加入VLAN1
DCRS-5650-28(config-Vlan1)#exit
DCRS-5650-28(config)#vlan 2
DCRS-5650-28(config-Vlan2)#switchport interface e0/0/11-20
                                         //端口11～20加入VLAN2
DCRS-5650-28(config-Vlan2)#exit
```

2. 给 VLAN 设置 IP 地址。

```
DCRS-5650-28(config)#interface vlan 1
DCRS-5650-28(config-if-Vlan1)#ip address 192.168.1.1 255.255.255.0
                        //设置VLAN1的IP地址为192.168.1.1
DCRS-5650-28(config-if-Vlan1)#exit
DCRS-5650-28(config)#interface vlan 2
DCRS-5650-28(config-if-Vlan2)#ip address 192.168.2.1 255.255.255.0
                        //设置VLAN2的IP地址为192.168.2.1
DCRS-5650-28(config-if-Vlan2)#exit
```

3. 查看路由表。

```
DCRS-5650-28(config)#show ip route
Codes: K - kernel, C - connected, S - static, R - RIP, B - BGP
       O - OSPF, IA - OSPF inter area
       N1 - OSPF NSSA external type 1, N2 - OSPF NSSA external type 2
       E1 - OSPF external type 1, E2 - OSPF external type 2
       i - IS-IS, L1 - IS-IS level-1, L2 - IS-IS level-2, ia - IS-IS inter area
       * - candidate default
C    127.0.0.0/8 is directly connected, Loopback
C    192.168.1.0/24 is directly connected, Vlan1
C    192.168.2.0/24 is directly connected, Vlan2
Total routes are: 3 item(s)
```

可以看到此交换机有两条直连路由 192.168.1.0/24 和 192.168.2.0/24。

4. 查看交换机配置。

```
DCRS-5650-28(config)#show running-config
no service password-encryption
hostname DCRS-5650-28
vendorlocation China
vendorContact 800-810-9119
vlan 1
vlan 2
```

```
Interface Ethernet0/0/1
Interface Ethernet0/0/2
略
Interface Ethernet0/0/10
Interface Ethernet0/0/11
 switchport access vlan 2
Interface Ethernet0/0/12
 switchport access vlan 2
Interface Ethernet0/0/13
 switchport access vlan 2
Interface Ethernet0/0/14
 switchport access vlan 2
Interface Ethernet0/0/15
 switchport access vlan 2
Interface Ethernet0/0/16
 switchport access vlan 2
Interface Ethernet0/0/17
 switchport access vlan 2
Interface Ethernet0/0/18
 switchport access vlan 2
Interface Ethernet0/0/19
 switchport access vlan 2
Interface Ethernet0/0/20
 switchport access vlan 2
Interface Ethernet0/0/21
Interface Ethernet0/0/22
Interface Ethernet0/0/23
Interface Ethernet0/0/24
Interface Ethernet0/0/25
Interface Ethernet0/0/26
Interface Ethernet0/0/27
Interface Ethernet0/0/28
interface Vlan1
ip address 192.168.1.1 255.255.255.0    //VLAN1 的 IP 地址为 192.168.1.1
interface Vlan2
 ip address 192.168.2.1 255.255.255.0   //VLAN2 的 IP 地址为 192.168.2.1
no login
end
```

5. 显示 VLAN 中包含的端口，Show VLAN。

```
DCRS-5650-28(config)#show vlan
VLAN Name            Type     Media    Ports
-----------------------------------------------------------------
1    default         Static   ENET     Ethernet0/0/1     Ethernet0/0/2
                                       Ethernet0/0/3     Ethernet0/0/4
                                       Ethernet0/0/5     Ethernet0/0/6
                                       Ethernet0/0/7     Ethernet0/0/8
```

				Ethernet0/0/9	Ethernet0/0/10
				Ethernet0/0/21	Ethernet0/0/22
				Ethernet0/0/23	Ethernet0/0/24
				Ethernet0/0/25	Ethernet0/0/26
				Ethernet0/0/27	Ethernet0/0/28
2	VLAN0002	Static	ENET	Ethernet0/0/11	Ethernet0/0/12
				Ethernet0/0/13	Ethernet0/0/14
				Ethernet0/0/15	Ethernet0/0/16
				Ethernet0/0/17	Ethernet0/0/18
				Ethernet0/0/19	Ethernet0/0/20

3.2 实训二 使用三层交换机实现二层交换机 VLAN 间的路由

任务描述

某公司网络中有 3 台交换机，一台 DCRS-5650 作为核心交换机，两台 DCS3950 作为接入交换机。其中 DCRS-5650 的 23 端口连接二层交换机 DCSA 的 23 端口，DCRS-5650 的 24 端口连接二层交换机 DCSB 的 24 端口。交换机间链路均设置为 Trunk 端口。由于业务需要所有交换机中有两个 VLAN，分别为 VLAN1 和 VLAN2。DCRS-5650 的 VLAN1 的 IP 地址为 "192.168.1.1/24"，VLAN2 的 IP 地址为 "192.168.2.1/24"。DCSA 和 DCSB 中 1~10 端口属于 VLAN1，11~20 端口属于 VLAN2。

网络拓扑图如图 3-2 所示。

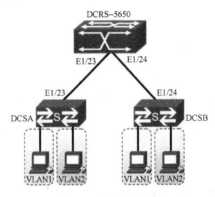

图 3-2 网络拓扑图

任务准备

三层交换机和二层交换机在 VLAN 划分、接口加入 VLAN 方式、VLAN 允许原理方面没有任何差别，只不过三层交换机多了路由功能，客户机在二层交换机和三层交换机混合的网络中只要设置了正确的网关，也就能正常通信。在配置过程中注意：在多 VLAN 的环境中，交换机间链路一般情况下应该设置为 Trunk 端口。

任务实施

DCRS-5650

1. 创建相应 VLAN，添加相应端口，设置相应 IP。

```
DCRS-5650-28>enable
DCRS-5650-28#config
DCRS-5650-28(config)#vlan 1
DCRS-5650-28(Config-Vlan1)#switchport interface e0/0/1-10
//端口 1～10 加入 VLAN1
DCRS-5650-28(Config-Vlan1)#exit
DCRS-5650-28(config)#vlan 2
DCRS-5650-28(Config-Vlan2)#switchport interface e0/0/11-20
//端口 11～20 加入 VLAN2
DCRS-5650-28(Config-Vlan2)#exit
DCRS-5650-28(config)#interface vlan 1
DCRS-5650-28(Config-if-Vlan1)#ip address 192.168.1.1 255.255.255.0
//设置 VLAN1 的 IP 地址为 192.168.1.1
DCRS-5650-28(Config-if-Vlan1)#exit
DCRS-5650-28(config)#interface vlan 2
DCRS-5650-28(Config-if-Vlan2)#ip address 192.168.2.1 255.255.255.0
//设置 VLAN2 的 IP 地址为 192.168.2.1
DCRS-5650-28(Config-if-Vlan2)#exit
```

2. 设置 23、24 端口为 Trunk 端口，并允许所有 VLAN 通过。

```
DCRS-5650-28(config)#interface e0/0/23-24            //进入端口 Range
DCRS-5650-28(Config-If-Port-Range)#switchport mode trunk
//设置端口类型为 Trunk
Set the port Ethernet0/0/23 mode TRUNK successfully
Set the port Ethernet0/0/24 mode TRUNK successfully
DCRS-5650-28(Config-If-Port-Range)#switchport trunk allowed vlan all
//设置允许所有 VLAN 通过 Trunk 端口
set the port Ethernet0/0/23 allowed vlan successfully
set the port Ethernet0/0/24 allowed vlan successfully
DCRS-5650-28(Config-If-Port-Range)#exit
```

3. 查看交换机配置。

```
DCRS-5650-28(config)#show running-config
no service password-encryption
hostname DCRS-5650-28
vendorlocation China
vendorContact 800-810-9119
vlan 1
vlan 2
Interface Ethernet0/0/1
Interface Ethernet0/0/2
略
```

```
Interface Ethernet0/0/10
Interface Ethernet0/0/11
 switchport access vlan 2
Interface Ethernet0/0/12
 switchport access vlan 2
Interface Ethernet0/0/13
 switchport access vlan 2
Interface Ethernet0/0/14
 switchport access vlan 2
Interface Ethernet0/0/15
 switchport access vlan 2
Interface Ethernet0/0/16
 switchport access vlan 2
Interface Ethernet0/0/17
 switchport access vlan 2
Interface Ethernet0/0/18
 switchport access vlan 2
Interface Ethernet0/0/19
 switchport access vlan 2
Interface Ethernet0/0/20
 switchport access vlan 2
Interface Ethernet0/0/21
Interface Ethernet0/0/22
Interface Ethernet0/0/23
 switchport mode trunk        //端口类型为Trunk，并允许所有VLAN数据通过
Interface Ethernet0/0/24
 switchport mode trunk        //端口类型为Trunk，并允许所有VLAN数据通过
Interface Ethernet0/0/25
Interface Ethernet0/0/26
Interface Ethernet0/0/27
Interface Ethernet0/0/28
interface Vlan1
 ip address 192.168.1.1 255.255.255.0   //VLAN1的IP地址为192.168.1.1
interface Vlan2
 ip address 192.168.2.1 255.255.255.0   //VLAN2的IP地址为192.168.2.1
no login
end
```

4. 查看VLAN列表。

```
DCRS-5650-28(config)#show vlan
VLAN Name         Type      Media    Ports
----------------------------------------------------------------
1    default      Static    ENET     Ethernet0/0/1     Ethernet0/0/2
                                     Ethernet0/0/3     Ethernet0/0/4
                                     Ethernet0/0/5     Ethernet0/0/6
                                     Ethernet0/0/7     Ethernet0/0/8
                                     Ethernet0/0/9     Ethernet0/0/10
```

				Ethernet0/0/21	Ethernet0/0/22
				Ethernet0/0/23(T)	Ethernet0/0/24(T)
				Ethernet0/0/25	Ethernet0/0/26
				Ethernet0/0/27	Ethernet0/0/28
2	VLAN0002	Static	ENET	Ethernet0/0/11	Ethernet0/0/12
				Ethernet0/0/13	Ethernet0/0/14
				Ethernet0/0/15	Ethernet0/0/16
				Ethernet0/0/17	Ethernet0/0/18
				Ethernet0/0/19	Ethernet0/0/20
				Ethernet0/0/23(T)	Ethernet0/0/24(T)

配置完成后可以从 VLAN 列表中看到，端口 23 和端口 24 同时属于 VLAN1 和 VLAN2。

DCSA

1. 创建相应 VLAN，添加相应端口。

```
DCS-3950-28C>enable
DCS-3950-28C#config
DCS-3950-28C(config)#hostname DCSA
DCSA(config)#vlan1
DCSA(config-vlan1)#switchport interface e1/1-10
//端口 1～10 加入 VLAN 1
DCSA(config-vlan1)#exit
DCSA(config)#vlan2
DCSA(config-vlan2)#switchport interface e1/11-20
//端口 11～20 加入 VLAN2
DCSA(config-vlan2)#exit
```

2. 设置 23 端口为 Trunk 端口，并允许所有 VLAN 通过。

```
DCSA(config)#interface e1/23
DCSA(config-if-ethernet1/23)#switchport mode trunk
//设置端口类型为 Trunk
Set the port Ethernet1/23 mode Trunk successfully
DCSA(config-if-ethernet1/23)#switchport trunk allowed vlan all
//设置允许所有 VLAN 通过
DCSA(config-if-ethernet1/23)#exit
```

3. 查看交换机配置。

```
DCSA(config)#show running-config
no service password-encryption
hostname DCSA
sysLocation China
sysContact 800-810-9119
username admin privilege 15 password 0 admin
vlan 1-2
Interface Ethernet1/1
Interface Ethernet1/2
略
```

```
Interface Ethernet1/10
Interface Ethernet1/11
 switchport access vlan 2
Interface Ethernet1/12
 switchport access vlan
Interface Ethernet1/13
 switchport access vlan 2
Interface Ethernet1/14
 switchport access vlan 2
Interface Ethernet1/15
 switchport access vlan 2
Interface Ethernet1/16
 switchport access vlan 2
Interface Ethernet1/17
 switchport access vlan 2
Interface Ethernet1/18
 switchport access vlan 2
Interface Ethernet1/19
 switchport access vlan 2
Interface Ethernet1/20
 switchport access vlan 2
Interface Ethernet1/21
Interface Ethernet1/22
Interface Ethernet1/23
 switchport mode trunk              //端口类型为Trunk，并允许所有VLAN数据通过
Interface Ethernet1/24
Interface Ethernet1/25
Interface Ethernet1/26
Interface Ethernet1/27
Interface Ethernet1/28
no login
End
```

4. 查看 VLAN。

```
DCSA(config)#show vlan
VLAN Name          Type      Media   Ports
-----------------------------------------------------------------
1    default       Static    ENET    Ethernet1/1        Ethernet1/2
                                     Ethernet1/3        Ethernet1/4
                                     Ethernet1/5        Ethernet1/6
                                     Ethernet1/7        Ethernet1/8
                                     Ethernet1/9        Ethernet1/10
                                     Ethernet1/21       Ethernet1/22
                                     Ethernet1/23       Ethernet1/24
                                     Ethernet1/25       Ethernet1/26
                                     Ethernet1/27       Ethernet1/28
2    VLAN0002      Static    ENET    Ethernet1/11       Ethernet1/12
```

```
                            Ethernet1/13        Ethernet1/14
                            Ethernet1/15        Ethernet1/16
                            Ethernet1/17        Ethernet1/18
                            Ethernet1/19        Ethernet1/20
                            Ethernet1/23(T)
```

配置完成后可以从 VLAN 列表中看到,端口 23 同时属于 VLAN1 和 VLAN2。

DCSB

1. 创建相应 VLAN,添加相应端口。

```
DCS-3950-28C>enable
DCS-3950-28C#config
DCS-3950-28C(config)#hostname DCSB
DCSB(config)#vlan 1
DCSB(config-vlan1)#switchport interface e1/1-10
//端口 1~10 加入 VLAN1
DCSB(config-vlan1)#exit
DCSB(config)#vlan 2
DCSB(config-vlan2)#switchport interface e1/11-20
//端口 11~20 加入 VLAN2
DCSB(config-vlan2)#exit
```

2. 设置 24 端口为 Trunk 端口,并允许所有 VLAN 通过。

```
DCSB(config)#interface e1/24
DCSB(config-if-ethernet1/24)#switchport mode trunk
Set the port Ethernet1/24 mode Trunk successfully
DCSB(config-if-ethernet1/24)#switchport trunk allowed vlan all
//设置允许所有 VLAN 通过
DCSB(config-if-ethernet1/24)#exit
```

3. 查看交换机配置。

```
DCSB(config)#show running-config
no service password-encryption
hostname DCSB
sysLocation China
sysContact 800-810-9119
username admin privilege 15 password 0 admin
vlan 1-2
Interface Ethernet1/1
Interface Ethernet1/2
略
Interface Ethernet1/10
Interface Ethernet1/11
 switchport access vlan 2
Interface Ethernet1/12
 switchport access vlan 2
```

```
Interface Ethernet1/13
 switchport access vlan 2
Interface Ethernet1/14
 switchport access vlan 2
Interface Ethernet1/15
 switchport access vlan 2
Interface Ethernet1/16
 switchport access vlan 2
Interface Ethernet1/17
 switchport access vlan 2
Interface Ethernet1/18
 switchport access vlan 2
Interface Ethernet1/19
 switchport access vlan 2
Interface Ethernet1/20
 switchport access vlan 2
Interface Ethernet1/21
Interface Ethernet1/22
Interface Ethernet1/23
Interface Ethernet1/24
 switchport mode trunk              //端口类型为Trunk
Interface Ethernet1/25
Interface Ethernet1/26
Interface Ethernet1/27
Interface Ethernet1/28
no login
End
```

4. 查看 VLAN。

```
DCSB(config)#show vlan
VLAN Name          Type      Media   Ports
-----------------------------------------------------------------
1    default       Static    ENET    Ethernet1/1      Ethernet1/2
                                     Ethernet1/3      Ethernet1/4
                                     Ethernet1/5      Ethernet1/6
                                     Ethernet1/7      Ethernet1/8
                                     Ethernet1/9      Ethernet1/10
                                     Ethernet1/21     Ethernet1/22
                                     Ethernet1/23     Ethernet1/24
                                     Ethernet1/25     Ethernet1/26
                                     Ethernet1/27     Ethernet1/28
2    VLAN0002      Static    ENET    Ethernet1/11     Ethernet1/12
                                     Ethernet1/13     Ethernet1/14
                                     Ethernet1/15     Ethernet1/16
                                     Ethernet1/17     Ethernet1/18
                                     Ethernet1/19     Ethernet1/20
                                     Ethernet1/24(T)
```

配置完成后可以从 VLAN 列表中看到，端口 24 同时属于 VLAN1 和 VLAN2。

所有交换机配置完成后，DCRS-5650 默认会启动路由转发，同时把 VLAN1 中 PC 的 IP 地址设置为 192.168.1.0/24 网段，网关设置为 192.168.1.1，把 VLAN2 中 PC 的 IP 地址设置为 192.168.2.0/24 网段，网关设置为 192.168.2.1。这样两个网段内的 PC 就可以互相通信。

3.3　实训三　三层交换机静态路由配置

任务描述

某公司的核心网络由 3 台三层交换机组成，网络拓扑图如图 3-3 所示。DCRS-A：端口 1~10 属于 VLAN2，VLAN2 的 IP 地址为"192.168.2.2/24"；端口 11~20 属于 VLAN3，VLAN3 的 IP 地址为"192.168.3.1/24"。DSRS-B：端口 1~10 属于 VLAN1，VLAN1 的 IP 地址为"192.168.1.1/24"；端口 11~20 属于 VLAN2，VLAN2 的 IP 地址为"192.168.2.1/24"。DCRS-C：端口 1~10 属于 VLAN3，VLAN3 的 IP 地址为"192.168.3.2/24"；端口 11~20 属于 VLAN4，VLAN4 的 IP 地址为"192.168.4.1/24"。管理员决定用手工配置静态路由来建立网络的路由表。

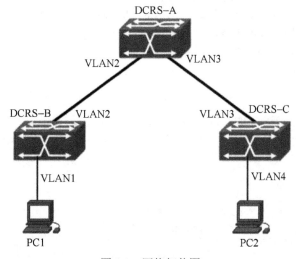

图 3-3　网络拓扑图

任务准备

静态路由：静态路由是指由网络管理员手工配置的路由信息。当网络的拓扑结构或链路的状态发生变化时，网络管理员需要手工去修改路由表中相关的静态路由信息。静态路由信息在默认情况下不会传递给其他的路由器。

默认路由：在路由表中，默认路由以目的网络为 0.0.0.0、子网掩码为 0.0.0.0 的形式出现。默认路由是一条特殊的静态路由，是在没有找到匹配的路由时使用的路由。如果数据包的目的地址不能与任何路由相匹配，那么系统将使用默认路由转发该数据包。

任务实施

DCRS-A

1. 创建相应 VLAN，添加相应端口，设置相应 IP 地址。

```
DCRS-5650-28>enable
DCRS-5650-28#config
DCRS-5650-28(config)#hostname DCRS-A
DCRS-A(config)#vlan 2
DCRS-A(Config-Vlan2)#switchport interface e0/0/1-10
//端口 1～10 加入 VLAN2
DCRS-A(Config-Vlan2)#exit
DCRS-A(config)#vlan 3
DCRS-A(config-Vlan3)#switchport interface e0/0/11-20
//端口 11～20 加入 VLAN3
DCRS-A(Config-Vlan3)#exit
DCRS-A(config)#interface vlan 2
DCRS-A(config-if-Vlan2)#ip address 192.168.2.2 255.255.255.0
//设置 VLAN2 的 IP 地址为 192.168.2.2
DCRS-A(config-if-Vlan2)#exit
DCRS-A(config)#interface vlan 3
DCRS-A(config-if-Vlan3)#ip address 192.168.3.1 255.255.255.0
//设置 VLAN3 的 IP 地址为 192.168.3.1
DCRS-A(config-if-Vlan3)#exit
```

2. 添加静态路由。

```
DCRS-A(config)#ip route 192.168.1.0 255.255.255.0 192.168.2.1
//添加到 192.168.1.0 网段的路由表项，下一跳为 192.168.2.1
DCRS-A(config)#ip route 192.168.4.0 255.255.255.0 192.168.3.2
//添加到 192.168.4.0 网段的路由表项，下一跳为 192.168.3.2
```

3. 查看交换机配置。

```
DCRS-A(config)#show running-config
no service password-encryption
hostname DCRS-A
vlan 1
vlan 2
vlan 3
Interface Ethernet0/0/1
 switchport access vlan 2                    //端口 1 属于 VLAN2
 Interface Ethernet0/0/2
 switchport access vlan 2                    //端口 2 属于 VLAN2
Interface Ethernet0/0/3
 switchport access vlan 2
Interface Ethernet0/0/4
 switchport access vlan 2
```

```
Interface Ethernet0/0/5
 switchport access vlan 2
Interface Ethernet0/0/6
 switchport access vlan 2
Interface Ethernet0/0/7
 switchport access vlan 2
Interface Ethernet0/0/8
 switchport access vlan 2
Interface Ethernet0/0/9
 switchport access vlan 2
Interface Ethernet0/0/10
 switchport access vlan 2
Interface Ethernet0/0/11                          //端口 11 属于 VLAN3
 switchport access vlan 3
Interface Ethernet0/0/12
 switchport access vlan 3                         //端口 12 属于 VLAN3
Interface Ethernet0/0/13
 switchport access vlan 3
Interface Ethernet0/0/14
 switchport access vlan 3
Interface Ethernet0/0/15
 switchport access vlan 3
Interface Ethernet0/0/16
 switchport access vlan 3
Interface Ethernet0/0/17
 switchport access vlan 3
Interface Ethernet0/0/18
 switchport access vlan 3
Interface Ethernet0/0/19
 switchport access vlan 3
Interface Ethernet0/0/20
 switchport access vlan 3
Interface Ethernet0/0/21
Interface Ethernet0/0/22
Interface Ethernet0/0/23
Interface Ethernet0/0/24
Interface Ethernet0/0/25
Interface Ethernet0/0/26
Interface Ethernet0/0/27
Interface Ethernet0/0/28
interface Vlan2
 ip address 192.168.2.2 255.255.255.0   //接口 VLAN2 的 IP 地址为 192.168.2.2
interface Vlan3
 ip address 192.168.3.1 255.255.255.0   //接口 VLAN2 的 IP 地址为 192.168.3.1
ip route 192.168.1.0/24 192.168.2.1
//到 192.168.1.0 网段的路由表项,下一跳为 192.168.2.1
ip route 192.168.4.0/24 192.168.3.2
```

```
//到192.168.4.0网段的路由表项，下一跳为192.168.3.2
no login
end
```

4．查看路由表。

```
DCRS-A(config)#show ip route
Codes: K - kernel, C - connected, S - static, R - RIP, B - BGP
       O - OSPF, IA - OSPF inter area
       N1 - OSPF NSSA external type 1, N2 - OSPF NSSA external type 2
       E1 - OSPF external type 1, E2 - OSPF external type 2
       i - IS-IS, L1 - IS-IS level-1, L2 - IS-IS level-2, ia - IS-IS inter area
       * - candidate default
C      127.0.0.0/8 is directly connected, Loopback
S      192.168.1.0/24 [1/0] via 192.168.2.1, Vlan2
//到192.168.1.0网段的路由表项，下一跳为192.168.2.1
C      192.168.2.0/24 is directly connected, Vlan2
C      192.168.3.0/24 is directly connected, Vlan3
S      192.168.4.0/24 [1/0] via 192.168.3.2, Vlan3
//到192.168.4.0网段的路由表项，下一跳为192.168.3.2
```

DCRS-B

1．创建相应 VLAN，添加相应端口，设置相应 IP 地址。

```
DCRS-5650-28>enable
DCRS-5650-28#config
DCRS-5650-28(config)#hostname DCRS-B
DCRS-B(config)#vlan 1
DCRS-B(Config-Vlan1)#switchport interface e0/0/1-10
//端口 1～10 加入 VLAN1
DCRS-B(Config-Vlan1)#exit
DCRS-B(config)#vlan 2
DCRS-B(config-Vlan2)#switchport interface e0/0/11-20
//端口 11～20 加入 VLAN2
DCRS-B(Config-Vlan2)#exit
DCRS-B(config)#interface vlan 1
DCRS-B(config-if-Vlan1)#ip address 192.168.1.1 255.255.255.0
//设置 VLAN1 的 IP 地址为 192.168.1.1
DCRS-B(config-if-Vlan1)#exit
DCRS-B(config)#interface vlan 2
DCRS-B(config-if-Vlan2)#ip address 192.168.2.1 255.255.255.0
//设置 VLAN2 的 IP 地址为 192.168.2.1
DCRS-B(config-if-Vlan2)#exit
```

2．配置静态路由。

```
DCRS-B(config)#ip route 0.0.0.0 0.0.0.0 192.168.2.2
```

//添加到默认路由表项，下一跳为192.168.2.2

3. 查看交换机配置。

```
DCRS-B(config)#show running-config
no service password-encryption
hostname DCRS-B
vendorlocation China
vendorContact 800-810-9119
vlan 1
vlan 2
Interface Ethernet0/0/1
Interface Ethernet0/0/2
Interface Ethernet0/0/3
Interface Ethernet0/0/4
Interface Ethernet0/0/5
Interface Ethernet0/0/6
Interface Ethernet0/0/7
Interface Ethernet0/0/8
Interface Ethernet0/0/9
Interface Ethernet0/0/10
Interface Ethernet0/0/11
 switchport access vlan 2                //端口1属于VLAN2
Interface Ethernet0/0/12
 switchport access vlan 2
Interface Ethernet0/0/13
 switchport access vlan 2
Interface Ethernet0/0/14
 switchport access vlan 2
Interface Ethernet0/0/15
 switchport access vlan 2!
Interface Ethernet0/0/16
 switchport access vlan 2
Interface Ethernet0/0/17
 switchport access vlan 2
Interface Ethernet0/0/18
 switchport access vlan 2
Interface Ethernet0/0/19
 switchport access vlan 2
Interface Ethernet0/0/20
 switchport access vlan 2
Interface Ethernet0/0/21
Interface Ethernet0/0/22
Interface Ethernet0/0/23
Interface Ethernet0/0/24
Interface Ethernet0/0/25
Interface Ethernet0/0/26
```

```
    Interface Ethernet0/0/27
    Interface Ethernet0/0/28
    interface Vlan1
     ip address 192.168.1.1 255.255.255.0   //接口 VLAN1 的 IP 地址为 192.168.1.1
    interface Vlan2
     ip address 192.168.2.1 255.255.255.0   //接口 VLAN2 的 IP 地址为 192.168.2.1
    ip route 0.0.0.0/0 192.168.2.2          //默认路由表项，下一跳为 192.168.2.2
    no login
    end
```

4. 查看路由表。

```
    DCRS-B(config)#show ip route
    Codes: K - kernel, C - connected, S - static, R - RIP, B - BGP
           O - OSPF, IA - OSPF inter area
           N1 - OSPF NSSA external type 1, N2 - OSPF NSSA external type 2
           E1 - OSPF external type 1, E2 - OSPF external type 2
           i - IS-IS, L1 - IS-IS level-1, L2 - IS-IS level-2, ia - IS-IS inter area
           * - candidate default
    Gateway of last resort is 192.168.2.2 to network 0.0.0.0
    S*     0.0.0.0/0 [1/0] via 192.168.2.2, Vlan2
//默认路由表项，下一跳为 192.168.2.2
    C      127.0.0.0/8 is directly connected, Loopback
    C      192.168.1.0/24 is directly connected, Vlan1
    C      192.168.2.0/24 is directly connected, Vlan2
    Total routes are : 4 item(s)
```

DCRS-C

1. 创建相应 VLAN，添加相应端口，设置相应 IP 地址。

```
    DCRS-5650-28>enable
    DCRS-5650-28#config
    DCRS-5650-28(config)#hostname DCRS-C
    DCRS-C(config)#vlan 3
    DCRS-C(Config-Vlan3)#switchport interface e0/0/1-10   //端口 1～10 加入
VLAN3
    DCRS-C(Config-Vlan3)#exit
    DCRS-C(config)#vlan 4
    DCRS-C(config-Vlan4)#switchport interface e0/0/11-20  //端口 11～20 加入
VLAN4
    DCRS-C(Config-Vlan4)#exit
    DCRS-C(config)#interface vlan 3
    DCRS-C(config-if-Vlan3)#ip address 192.168.3.2 255.255.255.0
    //接口 VLAN3 的 IP 地址为 192.168.3.2
    DCRS-C(config-if-Vlan3)#exit
    DCRS-C(config)#interface vlan 4
```

```
DCRS-C(config-if-Vlan4)#ip address 192.168.4.1 255.255.255.0
//接口VLAN4的IP地址为192.168.4.1
DCRS-C(config-if-Vlan4)#exit
```

2．设置静态路由。

```
DCRS-C(config)#ip route 0.0.0.0 0.0.0.0 192.168.3.1
//添加默认路由表项，下一跳为192.168.3.1
```

3．查看交换机配置。

```
DCRS-C(config)#show running-config
no service password-encryption
hostname DCRS-C
vlan 1
vlan 3
vlan 4
Interface Ethernet0/0/1
 switchport access vlan 3                                    //端口1属于VLAN3
Interface Ethernet0/0/2
 switchport access vlan 3
Interface Ethernet0/0/3
 switchport access vlan 3
Interface Ethernet0/0/4
 switchport access vlan 3
Interface Ethernet0/0/5
 switchport access vlan 3
Interface Ethernet0/0/6
 switchport access vlan 3
Interface Ethernet0/0/7
 switchport access vlan 3
Interface Ethernet0/0/8
 switchport access vlan 3
Interface Ethernet0/0/9
 switchport access vlan 3
Interface Ethernet0/0/10
 switchport access vlan 3
Interface Ethernet0/0/11
 switchport access vlan 4                                    //端口11属于VLAN4
Interface Ethernet0/0/12
 switchport access vlan 4
Interface Ethernet0/0/13
 switchport access vlan 4
Interface Ethernet0/0/14
 switchport access vlan 4
Interface Ethernet0/0/15
 switchport access vlan 4
Interface Ethernet0/0/16
```

```
  switchport access vlan 4
 Interface Ethernet0/0/17
  switchport access vlan 4
 Interface Ethernet0/0/18
  switchport access vlan 4
 Interface Ethernet0/0/19
  switchport access vlan 4
 Interface Ethernet0/0/20
  switchport access vlan 4
 Interface Ethernet0/0/21
 Interface Ethernet0/0/22
 Interface Ethernet0/0/23
 Interface Ethernet0/0/24
 Interface Ethernet0/0/25
 Interface Ethernet0/0/26
 Interface Ethernet0/0/27
 Interface Ethernet0/0/28
 interface Vlan3
  ip address 192.168.3.2 255.255.255.0    //接口 VLAN3 的 IP 地址为 192.168.3.2
 interface Vlan4
  ip address 192.168.4.1 255.255.255.0    //接口 VLAN4 的 IP 地址为 192.168.4.1
 ip route 0.0.0.0/0 192.168.3.1           //默认路由表项，下一跳为 192.168.3.1
 no login
 end
```

4．查看路由表。

```
DCRS-C(config)#show ip route
Codes: K - kernel, C - connected, S - static, R - RIP, B - BGP
       O - OSPF, IA - OSPF inter area
       N1 - OSPF NSSA external type 1, N2 - OSPF NSSA external type 2
       E1 - OSPF external type 1, E2 - OSPF external type 2
       i - IS-IS, L1 - IS-IS level-1, L2 - IS-IS level-2, ia - IS-IS inter area
       * - candidate default
Gateway of last resort is 192.168.3.1 to network 0.0.0.0
S*     0.0.0.0/0 [1/0] via 192.168.3.1, Vlan3
//默认路由表项，下一跳为 192.168.3.1
C      127.0.0.0/8 is directly connected, Loopback
C      192.168.3.0/24 is directly connected, Vlan3
C      192.168.4.0/24 is directly connected, Vlan4
```

经过上述对 3 个三层交换机的配置，每个交换机的路由表中都包含了内网所有网段或默认路由，内网所有用户之间实现了互联互通，可以通过 ping 测试。

注意：本实训中 DCRS-B 和 DCRS-C 两台交换机中仅设置了一条默认路由，这种设置必须在本三层交换机是末节交换机的情况下才能使用。

3.4 实训四 三层交换机 RIP 动态路由配置

任务描述

某公司的核心网络由 3 台三层交换机组成，网络拓扑图如图 3-4 所示。DCRS-A：端口 1~10 属于 VLAN2，VLAN2 的 IP 地址为"192.168.2.2/24"；端口 11~20 属于 VLAN3，VLAN3 的 IP 地址为"192.168.3.1/24"。DSRS-B：端口 1~10 属于 VLAN1，VLAN1 的 IP 地址为"192.168.1.1/24"；端口 11~20 属于 VLAN2，VLAN2 的 IP 地址为"192.168.2.1/24"。DCRS-C：端口 1~10 属于 VLAN3，VLAN3 的 IP 地址为"192.168.3.2/24"；端口 11~20 属于 VLAN4，VLAN4 的 IP 地址为"192.168.4.1/24"。管理员决定用 RIP 路由协议来建立网络的路由表。

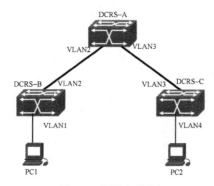

图 3-4 网络拓扑图

任务准备

RIP：RIP（Routing Information Protocol）即路由信息协议。RIP 是自治系统内部使用的协议，即内部网关协议。它使用的是距离矢量算法，使用 UDP 的 520 端口进行 RIP 进程之间的通信，每 30 秒会与相邻的路由器交换整个路由表，动态地建立路由表。RIP 协议以跳数作为网络度量值，网络直径不超过 15 跳，适合于中小型网络，16 跳时将被认为网络不可达。

RIP 主要有两个版本：RIPv1 和 RIPv2。RIP 协议采用广播或组播进行通信，其中 RIPv1 只支持广播，而 RIPv2 除支持广播外还支持组播。RIPv1 是有类路由协议，RIPv2 是无类路由协议，即 RIPv2 的报文中含有掩码信息。交换机中默认使用 RIPv2 版本。

任务实施

DCRS-A

1. 创建相应 VLAN，添加相应端口，设置相应 IP 地址。

```
DCRS-5650-28>enable
DCRS-5650-28#config
DCRS-5650-28(config)#hostname DCRS-A
```

```
DCRS-A(config)#vlan 2
DCRS-A(Config-Vlan2)#switchport interface e0/0/1-10          //端口 1～10
加入 VLAN2
DCRS-A(Config-Vlan2)#exit
DCRS-A(config)#vlan 3
DCRS-A(config-Vlan3)#switchport interface e0/0/11-20         //端口 11～20
加入 VLAN3
DCRS-A(Config-Vlan3)#exit
DCRS-A(config)#interface vlan 2
DCRS-A(config-if-Vlan2)#ip address 192.168.2.2 255.255.255.0
//设置 VLAN2 的 IP 地址为 192.168.2.2
DCRS-A(config-if-Vlan2)#exit
DCRS-A(config)#interface vlan 3
DCRS-A(config-if-Vlan3)#ip address 192.168.3.1 255.255.255.0
//设置 VLAN3 的 IP 地址为 192.168.3.1
DCRS-A(config-if-Vlan3)#exit
```

2. 配置 RIP 协议。

```
DCRS-A(config)#router rip                                    //进入 RIP 视图
DCRS-A(config-router)#version 2                              //设置为 RIPv2
DCRS-A(config-router)#network 192.168.2.0/24
//在 192.168.2.0 网段启用 RIP 路由协议
DCRS-A(config-router)#network 192.168.3.0/24
//在 192.168.3.0 网段启用 RIP 路由协议
```

3. 查看交换机配置。

```
DCRS-A(config)#show running-config
no service password-encryption
hostname DCRS-A
vlan 1
vlan 2
vlan 3
Interface Ethernet0/0/1
 switchport access vlan 2
Interface Ethernet0/0/2
 switchport access vlan 2
Interface Ethernet0/0/3
 switchport access vlan 2
Interface Ethernet0/0/4
 switchport access vlan 2
Interface Ethernet0/0/5
 switchport access vlan 2
Interface Ethernet0/0/6
 switchport access vlan 2
Interface Ethernet0/0/7
 switchport access vlan 2
Interface Ethernet0/0/8
```

```
 switchport access vlan 2
Interface Ethernet0/0/9
 switchport access vlan 2
Interface Ethernet0/0/10
 switchport access vlan 2
Interface Ethernet0/0/11
switchport access vlan 3
Interface Ethernet0/0/12
switchport access vlan 3
Interface Ethernet0/0/13
 switchport access vlan 3
Interface Ethernet0/0/14
 switchport access vlan 3
Interface Ethernet0/0/15
 switchport access vlan 3
Interface Ethernet0/0/16
 switchport access vlan 3
Interface Ethernet0/0/17
switchport access vlan 3
Interface Ethernet0/0/18
 switchport access vlan 3
Interface Ethernet0/0/19
 switchport access vlan 3
Interface Ethernet0/0/20
 switchport access vlan 3
Interface Ethernet0/0/21
Interface Ethernet0/0/22
Interface Ethernet0/0/23
Interface Ethernet0/0/24
Interface Ethernet0/0/25
Interface Ethernet0/0/26
Interface Ethernet0/0/27
Interface Ethernet0/0/28
interface Vlan2
 ip address 192.168.2.2 255.255.255.0    //接口 VLAN2 的 IP 地址为 192.168.2.2
interface Vlan3
 ip address 192.168.3.1 255.255.255.0    //接口 VLAN3 的 IP 地址为 192.168.3.1
router rip                               //启用 RIPv2
  network 192.168.2.0/24                 //192.168.2.0 网段启用 RIP 路由协议
  network 192.168.3.0/24                 //192.168.3.0 网段启用 RIP 路由协议
no login
End
```

DCRS-B

1. 创建相应 VLAN，添加相应端口，设置相应 IP 地址。

```
DCRS-5650-28>enable
```

```
DCRS-5650-28#config
DCRS-5650-28(config)#hostname DCRS-B
DCRS-B(config)#vlan 1
DCRS-B(Config-Vlan1)#switchport interface e0/0/1-10        //端口 1~10
加入 VLAN1
DCRS-B(Config-Vlan1)#exit
DCRS-B(config)#vlan 2
DCRS-B(config-Vlan2)#switchport interface e0/0/11-20       //端口 11~20
加入 VLAN2
DCRS-B(Config-Vlan2)#exit
DCRS-B(config)#interface vlan 1
DCRS-B(config-if-Vlan1)#ip address 192.168.1.1 255.255.255.0
//设置 VLAN1 的 IP 地址为 192.168.1.1
DCRS-B(config-if-Vlan1)#exit
DCRS-B(config)#interface vlan 2
DCRS-B(config-if-Vlan2)#ip address 192.168.2.1 255.255.255.0
//设置 VLAN2 的 IP 地址为 192.168.2.1
DCRS-B(config-if-Vlan2)#exit
```

2. 配置 RIP 协议。

```
DCRS-B(config)#router rip                          //进入 RIP 视图
DCRS-B(config-router)#version 2                    //启用 RIPv2 版本
DCRS-B(config-router)#network 192.168.1.0/24
//192.168.1.0 网段启用 RIP 路由协议
DCRS-B(config-router)#network 192.168.2.0/24
//192.168.2.0 网段启用 RIP 路由协议
```

3. 查看交换机配置。

```
DCRS-B(config)#show running-config
no service password-encryption
hostname DCRS-B
vendorlocation China
vendorContact 800-810-9119
vlan 1
vlan 2
Interface Ethernet0/0/1
Interface Ethernet0/0/2
Interface Ethernet0/0/3
Interface Ethernet0/0/4
Interface Ethernet0/0/5
Interface Ethernet0/0/6
Interface Ethernet0/0/7
nterface Ethernet0/0/8
Interface Ethernet0/0/9
Interface Ethernet0/0/10
Interface Ethernet0/0/11
 switchport access vlan 2
```

```
Interface Ethernet0/0/12
 switchport access vlan 2
Interface Ethernet0/0/13
 switchport access vlan 2
Interface Ethernet0/0/14
 switchport access vlan 2
Interface Ethernet0/0/15
 switchport access vlan 2
Interface Ethernet0/0/16
 switchport access vlan 2
Interface Ethernet0/0/17
 switchport access vlan 2
Interface Ethernet0/0/18
 switchport access vlan 2
Interface Ethernet0/0/19
 switchport access vlan 2
Interface Ethernet0/0/20
 switchport access vlan 2
Interface Ethernet0/0/21
Interface Ethernet0/0/22
Interface Ethernet0/0/23
Interface Ethernet0/0/24
Interface Ethernet0/0/25
Interface Ethernet0/0/26
Interface Ethernet0/0/27
Interface Ethernet0/0/28
interface Vlan1
 ip address 192.168.1.1 255.255.255.0      //接口VLAN1的IP地址为192.168.1.1
interface Vlan2
 ip address 192.168.2.1 255.255.255.0      //接口VLAN2的IP地址为192.168.2.1
router rip                                 //启用RIPv2版本
 network 192.168.1.0/24                    //192.168.1.0网段启用RIP路由协议
 network 192.168.2.0/24                    //192.168.2.0网段启用RIP路由协议
no login
end
```

DCRS-C

1. 创建相应 VLAN，添加相应端口，设置相应 IP 地址。

```
DCRS-5650-28>enable
DCRS-5650-28#config
DCRS-5650-28(config)#hostname DCRS-C
DCRS-C(config)#vlan 3
DCRS-C(Config-Vlan3)#switchport interface e0/0/1-10    //端口1~10加入VLAN3
DCRS-C(Config-Vlan3)#exit
DCRS-C(config)#vlan 4
DCRS-C(config-Vlan4)#switchport interface e0/0/11-20   //端口11~20加入VLAN4
```

```
DCRS-C(Config-Vlan4)#exit
DCRS-C(config)#interface vlan 3
DCRS-C(config-if-Vlan3)#ip address 192.168.3.2 255.255.255.0
//接口 VLAN3 的 IP 地址为 192.168.3.2
DCRS-C(config-if-Vlan3)#exit
DCRS-C(config)#interface vlan 4
DCRS-C(config-if-Vlan4)#ip address 192.168.4.1 255.255.255.0
//接口 VLAN4 的 IP 地址为 192.168.4.1
DCRS-C(config-if-Vlan4)#exit
```

2. 配置 RIP 协议。

```
DCRS-C(config)#router rip                           //进入 RIP 视图模式
DCRS-C(config-router)#version 2                     //启用 RIPv2
DCRS-C(config-router)#network 192.168.3.0/24  //192.168.3.0 网段启用 RIP
路由协议
DCRS-C(config-router)#network 192.168.4.0/24  //192.168.4.0 网段启用 RIP
路由协议
```

3. 查看交换机配置。

```
DCRS-C(config)#show running-config
no service password-encryption
hostname DCRS-C
vlan 1
vlan 3
vlan 4
Interface Ethernet0/0/1
 switchport access vlan 3
Interface Ethernet0/0/2
 switchport access vlan 3
Interface Ethernet0/0/3
 switchport access vlan 3
Interface Ethernet0/0/4
 switchport access vlan 3
Interface Ethernet0/0/5
 switchport access vlan 3
Interface Ethernet0/0/6
 switchport access vlan 3
Interface Ethernet0/0/7
 switchport access vlan 3
Interface Ethernet0/0/8
 switchport access vlan 3
Interface Ethernet0/0/9
 switchport access vlan 3
Interface Ethernet0/0/10
 switchport access vlan 3
Interface Ethernet0/0/11
 switchport access vlan 4
```

```
Interface Ethernet0/0/12
 switchport access vlan 4
Interface Ethernet0/0/13
 switchport access vlan 4
Interface Ethernet0/0/14
 switchport access vlan 4
Interface Ethernet0/0/15
 switchport access vlan 4
Interface Ethernet0/0/16
 switchport access vlan 4
Interface Ethernet0/0/17
 switchport access vlan 4
Interface Ethernet0/0/18
 switchport access vlan 4
Interface Ethernet0/0/19
 switchport access vlan 4
Interface Ethernet0/0/20
 switchport access vlan 4
Interface Ethernet0/0/21
Interface Ethernet0/0/22
Interface Ethernet0/0/23
Interface Ethernet0/0/24
Interface Ethernet0/0/25
Interface Ethernet0/0/26
Interface Ethernet0/0/27
Interface Ethernet0/0/28
interface Vlan3
 ip address 192.168.3.2 255.255.255.0    //接口VLAN3的IP地址为192.168.3.2
interface Vlan4
 ip address 192.168.4.1 255.255.255.0    //接口VLAN4的IP地址为192.168.4.1
router rip                               //启用RIPv2
 network 192.168.3.0/24                  //192.168.3.0网段启用RIP路由协议
 network 192.168.4.0/24                  //192.168.4.0网段启用RIP路由协议
no login
end
```

所有配置完成后查看路由器中的路由表项。

DCRS-A

```
DCRS-A(config)#show ip route
Codes: K - kernel, C - connected, S - static, R - RIP, B - BGP
       O - OSPF, IA - OSPF inter area
       N1 - OSPF NSSA external type 1, N2 - OSPF NSSA external type 2
       E1 - OSPF external type 1, E2 - OSPF external type 2
       i - IS-IS, L1 - IS-IS level-1, L2 - IS-IS level-2, ia - IS-IS inter
area
```

```
         * - candidate default
  C      127.0.0.0/8 is directly connected, Loopback
  R      192.168.1.0/24 [120/2] via 192.168.2.1, Vlan2, 00:48:11
//RIP 路由协议获得 192.168.1.0 网段的路由表项
  C      192.168.2.0/24 is directly connected, Vlan2
  C      192.168.3.0/24 is directly connected, Vlan3
  R      192.168.4.0/24 [120/2] via 192.168.3.2, Vlan3, 00:39:47
//RIP 路由协议获得 192.168.4.0 网段的路由表项
```

DCRS-B

```
DCRS-B(config)#show ip route
Codes: K - kernel, C - connected, S - static, R - RIP, B - BGP
       O - OSPF, IA - OSPF inter area
       N1 - OSPF NSSA external type 1, N2 - OSPF NSSA external type 2
       E1 - OSPF external type 1, E2 - OSPF external type 2
       i - IS-IS, L1 - IS-IS level-1, L2 - IS-IS level-2, ia - IS-IS inter area
       * - candidate default
  C      127.0.0.0/8 is directly connected, Loopback
  C      192.168.1.0/24 is directly connected, Vlan1
  C      192.168.2.0/24 is directly connected, Vlan2
  R      192.168.3.0/24 [120/2] via 192.168.2.2, Vlan2, 00:13:08
//RIP 路由协议获得 192.168.3.0 网段的路由表项
  R      192.168.4.0/24 [120/3] via 192.168.2.2, Vlan2, 00:04:02
//RIP 路由协议获得 192.168.4.0 网段的路由表项
```

DCRS-C

```
DCRS-C(config)#show ip route
Codes: K - kernel, C - connected, S - static, R - RIP, B - BGP
       O - OSPF, IA - OSPF inter area
       N1 - OSPF NSSA external type 1, N2 - OSPF NSSA external type 2
       E1 - OSPF external type 1, E2 - OSPF external type 2
       i - IS-IS, L1 - IS-IS level-1, L2 - IS-IS level-2, ia - IS-IS inter area
       * - candidate default
  C      127.0.0.0/8 is directly connected, Loopback
  R      192.168.1.0/24 [120/3] via 192.168.3.1, Vlan3, 00:01:55
//RIP 路由协议获得 192.168.1.0 网段的路由表项
  R      192.168.2.0/24 [120/2] via 192.168.3.1, Vlan3, 00:01:55
//RIP 路由协议获得 192.168.2.0 网段的路由表项
  C      192.168.3.0/24 is directly connected, Vlan3
  C      192.168.4.0/24 is directly connected, Vlan4
```

实训完成后，利用"show ip route"命令，可以看到 3 台交换机的路由表项，通过直连路由和 RIP 协议获得了内网所有的路由表项（192.168.1.0/24、192.168.2.0/24、192.168.3.0/24 和 192.168.4.0/24）。其中类型 C 为直连路由，类型 R 为从 RIP 路由协议获取的路由表项。

3.5 实训五 三层交换机 OSPF 动态路由配置

任务描述

某公司的核心网络由 3 台三层交换机组成,网络拓扑图如图 3-5 所示。DCRS-A:端口 1~10 属于 VLAN2,VLAN2 的 IP 地址为"192.168.2.2/24";端口 11~20 属于 VLAN3,VLAN3 的 IP 地址为"192.168.3.1/24"。DSRS-B:端口 1~10 属于 VLAN1,VLAN1 的 IP 地址为"192.168.1.1/24";端口 11~20 属于 VLAN2,VLAN2 的 IP 地址为"192.168.2.1/24"。DCRS-C:端口 1~10 属于 VLAN3,VLAN3 的 IP 地址为"192.168.3.2/24";端口 11~20 属于 VLAN4,VLAN4 的 IP 地址为"192.168.4.1/24"。管理员决定用 OSPF 路由协议来建立网络的路由表。

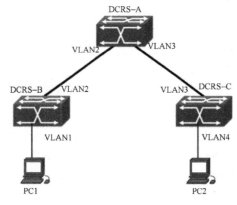

图 3-5 网络拓扑图

任务准备

OSPF:OSPF(Open Shortest Path First)即开放式最短路径优先,是一个内部网关路由协议,用于在单一自治系统(Autonomous System, AS)内决策路由。OSPF 是链路状态协议,而 RIP 是距离矢量协议。OSPF 通过路由器之间通告网络端口的状态来建立链路状态数据库,同一 AS 中所有的 OSPF 路由器都维护一个相同的描述这个 AS 结构的数据库,该数据库中存放的是路由域中相应链路的状态信息,OSPF 路由器通过这个数据库生成最短路径树,每个 OSPF 路由器使用这些最短路径构造路由表。

Router-ID:每一台 OSPF 路由器只有一个 Router-ID,Router-ID 使用 IP 地址的形式来表示,确定 Router-ID 的方法为:

①手工指定 Router-ID。

②路由器上活动 Loopback 端口中 IP 地址最大的,也就是数字最大的,如 C 类地址优先于 B 类地址,一个非活动的端口的 IP 地址是不能被选为 Router-ID 的。

③如果没有活动的 Loopback 端口,则选择活动物理端口 IP 地址最大的。

Router-ID 只在 OSPF 启动时计算,或者重置 OSPF 进程后计算。

COST:OSPF 使用端口的带宽来计算 Metric。

链路(Link):链路就是路由器上的端口,指运行在 OSPF 进程下的端口。

链路状态（Link-State）：链路状态（LSA）就是OSPF端口上的描述信息，例如端口上的IP地址、子网掩码、网络类型、Cost值等。OSPF路由器之间交换的并不是路由表，而是链路状态（LSA），OSPF通过获得网络中所有的链路状态信息，从而计算出到达每个目标精确的网络路径。

任务实施

DCRS-A

1. 创建相应VLAN，添加相应端口，设置相应IP地址。

```
DCRS-5650-28>enable
DCRS-5650-28#config
DCRS-5650-28(config)#hostname DCRS-A
DCRS-A(config)#vlan 2
DCRS-A(Config-Vlan2)#switchport interface e0/0/1-10        //端口1~10加入VLAN2
DCRS-A(Config-Vlan2)#exit
DCRS-A(config)#vlan 3
DCRS-A(Config-Vlan3)#switchport interface e0/0/11-20       //端口11~20加入VLAN3
DCRS-A(Config-Vlan3)#exit
DCRS-A(config)#interface vlan 2
DCRS-A(config-if-Vlan2)#ip address 192.168.2.2 255.255.255.0
//设置VLAN2的IP地址为192.168.2.2
DCRS-A(config-if-Vlan2)#exit
DCRS-A(config)#interface vlan 3
DCRS-A(config-if-Vlan3)#ip address 192.168.3.1 255.255.255.0
//设置VLAN3的IP地址为192.168.3.1
DCRS-A(config-if-Vlan3)#exit
```

2. 配置OSPF协议。

```
DCRS-A(config)#router ospf                                 //启动OSPF路由协议
DCRS-A(config-router)#network 192.168.2.0/24 area 0
//192.168.2.0网段运行OSPF路由协议
DCRS-A(config-router)#network 192.168.3.0/24 area 0
//192.168.3.0网段运行OSPF路由协议
```

3. 查看交换机配置。

```
DCRS-A(config)#show running-config
no service password-encryption
hostname DCRS-A
vlan 1
vlan 2
vlan 3
Interface Ethernet0/0/1
 switchport access vlan 2
```

```
Interface Ethernet0/0/2
 switchport access vlan 2
Interface Ethernet0/0/3
 switchport access vlan 2
Interface Ethernet0/0/4
 switchport access vlan 2
Interface Ethernet0/0/5
 switchport access vlan 2
Interface Ethernet0/0/6
 switchport access vlan 2
Interface Ethernet0/0/7
 switchport access vlan 2
Interface Ethernet0/0/8
 switchport access vlan 2
Interface Ethernet0/0/9
 switchport access vlan 2
Interface Ethernet0/0/10
 switchport access vlan 2
Interface Ethernet0/0/11
 switchport access vlan 3
Interface Ethernet0/0/12
 switchport access vlan 3
Interface Ethernet0/0/13
 switchport access vlan 3
Interface Ethernet0/0/14
 switchport access vlan 3
Interface Ethernet0/0/15
 switchport access vlan 3
Interface Ethernet0/0/16
 switchport access vlan 3
Interface Ethernet0/0/17
 switchport access vlan 3
Interface Ethernet0/0/18
 switchport access vlan 3
Interface Ethernet0/0/19
 switchport access vlan 3
Interface Ethernet0/0/20
 switchport access vlan 3
Interface Ethernet0/0/21
Interface Ethernet0/0/22
Interface Ethernet0/0/23
Interface Ethernet0/0/24
Interface Ethernet0/0/25
Interface Ethernet0/0/26
Interface Ethernet0/0/27
Interface Ethernet0/0/28
interface Vlan2
```

```
   ip address 192.168.2.2 255.255.255.0
  interface Vlan
   ip address 192.168.3.1 255.255.255.
  router ospf                                    //启用OSPF路由协议
   network 192.168.2.0/24 area 0                 //192.168.2.0网段运行OSPF路由协议
   network 192.168.3.0/24 area 0                 //192.168.3.0网段运行OSPF路由协议
  no login
  end
```

DCRS-B

1. 创建相应 VLAN，添加相应端口，设置相应 IP 地址。

```
DCRS-5650-28>enable
DCRS-5650-28#config
DCRS-5650-28(config)#hostname DCRS-B
DCRS-B(config)#vlan 1
DCRS-B(Config-Vlan1)#switchport interface e0/0/1-10         //端口1~10加入VLAN1
DCRS-B(Config-Vlan1)#exit
DCRS-B(config)#vlan 2
DCRS-B(config-Vlan2)#switchport interface e0/0/11-20        //端口11~20加入VLAN2
DCRS-B(Config-Vlan2)#exit
DCRS-B(config)#interface vlan 1
DCRS-B(config-if-Vlan1)#ip address 192.168.1.1 255.255.255.0
//设置VLAN1的IP地址为192.168.1.1
DCRS-B(config-if-Vlan1)#exit
DCRS-B(config)#interface vlan 2
DCRS-B(config-if-Vlan2)#ip address 192.168.2.1 255.255.255.0
//设置VLAN2的IP地址为192.168.2.1
DCRS-B(config-if-Vlan2)#exit
```

2. 配置 OSPF 协议。

```
DCRS-B(config)#router ospf 1                                //启动OSPF路由协议
DCRS-B(config-router)#network 192.168.1.0/24 area 0
//192.168.1.0网段运行OSPF路由协议
DCRS-B(config-router)#network 192.168.2.0/24 area 0
//192.168.2.0网段运行OSPF路由协议
```

3. 查看交换机配置。

```
DCRS-B(config)#show running-config
no service password-encryption
hostname DCRS-B
vendorlocation China
vendorContact 800-810-9119
vlan 1
vlan 2
```

```
Interface Ethernet0/0/1
Interface Ethernet0/0/2
Interface Ethernet0/0/3
Interface Ethernet0/0/4
Interface Ethernet0/0/5
Interface Ethernet0/0/6
Interface Ethernet0/0/7
Interface Ethernet0/0/8
Interface Ethernet0/0/9
Interface Ethernet0/0/10
Interface Ethernet0/0/11
 switchport access vlan 2
Interface Ethernet0/0/12
 switchport access vlan 2
Interface Ethernet0/0/13
 switchport access vlan 2
Interface Ethernet0/0/14
 switchport access vlan 2
Interface Ethernet0/0/15
 switchport access vlan 2
Interface Ethernet0/0/16
 switchport access vlan 2
Interface Ethernet0/0/17
 switchport access vlan 2
Interface Ethernet0/0/18
 switchport access vlan 2
Interface Ethernet0/0/19
 switchport access vlan 2
Interface Ethernet0/0/20
 switchport access vlan 2
Interface Ethernet0/0/21
Interface Ethernet0/0/22
Interface Ethernet0/0/23
Interface Ethernet0/0/24
Interface Ethernet0/0/25
Interface Ethernet0/0/26
Interface Ethernet0/0/27
Interface Ethernet0/0/28
interface Vlan1
 ip address 192.168.1.1 255.255.255.0   //接口 VLAN1 的 IP 地址为 192.168.1.1
interface Vlan2
 ip address 192.168.2.1 255.255.255.0   //接口 VLAN2 的 IP 地址为 192.168.2.1
router ospf 1
 network 192.168.1.0/24 area 0          //192.168.1.0 网段运行 OSPF 路由协议
 network 192.168.2.0/24 area 0          //192.168.2.0 网段运行 OSPF 路由协议
no login
end
```

DCRS-C

1. 创建相应 VLAN，添加相应端口，设置相应 IP 地址。

```
DCRS-5650-28>enable
DCRS-5650-28#config
DCRS-5650-28(config)#hostname DCRS-C
DCRS-C(config)#vlan 3
DCRS-C(Config-Vlan3)#switchport interface e0/0/1-10    //端口1～10加入VLAN3
DCRS-C(Config-Vlan3)#exit
DCRS-C(config)#vlan 4
DCRS-C(config-Vlan4)#switchport interface e0/0/11-20   //端口11～20加入VLAN4
DCRS-C(Config-Vlan4)#exit
DCRS-C(config)#interface vlan 3
DCRS-C(config-if-Vlan3)#ip address 192.168.3.2 255.255.255.0
//接口VLAN3 的 IP 地址为 192.168.3.2
DCRS-C(config-if-Vlan3)#exit
DCRS-C(config)#interface vlan 4
DCRS-C(config-if-Vlan4)#ip address 192.168.4.1 255.255.255.0
//接口VLAN4 的 IP 地址为 192.168.4.1
DCRS-C(config-if-Vlan4)#exit
```

2. 配置 OSPF 协议。

```
DCRS-C(config)#router ospf 1                          //启动OSPF路由协议
DCRS-C(config-router)#network 192.168.3.0/24 area 0
//192.168.3.0 网段运行 OSPF 路由协议
DCRS-C(config-router)#network 192.168.4.0/24 area 0
//192.168.4.0 网段运行 OSPF 路由协议
```

3. 查看交换机配置。

```
DCRS-C(config)#show running-config
no service password-encryption
hostname DCRS-C
vlan 1
vlan 3
vlan 4
Interface Ethernet0/0/1
 switchport access vlan 3
Interface Ethernet0/0/2
 switchport access vlan 3
Interface Ethernet0/0/3
 switchport access vlan 3
Interface Ethernet0/0/4
 switchport access vlan 3
Interface Ethernet0/0/5
 switchport access vlan 3
Interface Ethernet0/0/6
```

```
 switchport access vlan 3
Interface Ethernet0/0/7
 switchport access vlan 3
Interface Ethernet0/0/8
 switchport access vlan 3
Interface Ethernet0/0/9
 switchport access vlan 3
Interface Ethernet0/0/10
 switchport access vlan 3
Interface Ethernet0/0/11
 switchport access vlan 4
Interface Ethernet0/0/12
 switchport access vlan 4
Interface Ethernet0/0/13
 switchport access vlan 4
Interface Ethernet0/0/14
 switchport access vlan 4
Interface Ethernet0/0/15
 switchport access vlan 4
Interface Ethernet0/0/16
 switchport access vlan 4
Interface Ethernet0/0/17
 switchport access vlan 4
Interface Ethernet0/0/18
 switchport access vlan 4
Interface Ethernet0/0/19
 switchport access vlan 4
Interface Ethernet0/0/20
 switchport access vlan 4
Interface Ethernet0/0/21
Interface Ethernet0/0/22
Interface Ethernet0/0/23
Interface Ethernet0/0/24
Interface Ethernet0/0/25
Interface Ethernet0/0/26
Interface Ethernet0/0/27
Interface Ethernet0/0/28
interface Vlan3
 ip address 192.168.3.2 255.255.255.0   //接口 VLAN3 的 IP 地址为 192.168.3.2
interface Vlan4
 ip address 192.168.4.1 255.255.255.0   //接口 VLAN4 的 IP 地址为 192.168.4.1
 router ospf 1                          //启动 OSPF 路由协议
 network 192.168.3.0/24 area 0          //192.168.3.0 网段运行 OSPF 路由协议
 network 192.168.4.0/24 area 0          //192.168.4.0 网段运行 OSPF 路由协议
no login
end
```

所有配置完成后查看路由器中路由表项。

DCRS-A

```
DCRS-A(config)#show ip route
Codes: K - kernel, C - connected, S - static, R - RIP, B - BGP
       O - OSPF, IA - OSPF inter area
       N1 - OSPF NSSA external type 1, N2 - OSPF NSSA external type 2
       E1 - OSPF external type 1, E2 - OSPF external type 2
       i - IS-IS, L1 - IS-IS level-1, L2 - IS-IS level-2, ia - IS-IS inter area
       * - candidate default
   C    127.0.0.0/8 is directly connected, Loopback
   O    192.168.1.0/24 [110/20] via 192.168.2.1, Vlan2, 00:06:51
//OSPF 路由协议获得 192.168.1.0 网段的路由表项
   C    192.168.2.0/24 is directly connected, Vlan2
   C    192.168.3.0/24 is directly connected, Vlan3
   O    192.168.4.0/24 [110/20] via 192.168.3.2, Vlan3, 00:02:01
//OSPF 路由协议获得 192.168.4.0 网段的路由表项
```

DCRS-B

```
DCRS-B(config)#show ip route
Codes: K - kernel, C - connected, S - static, R - RIP, B - BGP
       O - OSPF, IA - OSPF inter area
       N1 - OSPF NSSA external type 1, N2 - OSPF NSSA external type 2
       E1 - OSPF external type 1, E2 - OSPF external type 2
       i - IS-IS, L1 - IS-IS level-1, L2 - IS-IS level-2, ia - IS-IS inter area
       * - candidate default
   C    127.0.0.0/8 is directly connected, Loopback
   C    192.168.1.0/24 is directly connected, Vlan1
   C    192.168.2.0/24 is directly connected, Vlan2
   O    192.168.3.0/24 [110/20] via 192.168.2.2, Vlan2, 00:05:59
//OSPF 路由协议获得 192.168.3.0 网段的路由表项
   O    192.168.4.0/24 [110/30] via 192.168.2.2, Vlan2, 00:01:16
//OSPF 路由协议获得 192.168.3.0 网段的路由表项
Total routes are : 5 item(s)
```

DCRS-C

```
DCRS-C(config)#show ip route
Codes: K - kernel, C - connected, S - static, R - RIP, B - BGP
       O - OSPF, IA - OSPF inter area
       N1 - OSPF NSSA external type 1, N2 - OSPF NSSA external type 2
       E1 - OSPF external type 1, E2 - OSPF external type 2
       i - IS-IS, L1 - IS-IS level-1, L2 - IS-IS level-2, ia - IS-IS inter area
```

```
         * - candidate default
C       127.0.0.0/8 is directly connected, Loopback
O       192.168.1.0/24 [110/30] via 192.168.3.1, Vlan3, 00:00:36
//OSPF 路由协议获得 192.168.3.0 网段的路由表项
O       192.168.2.0/24 [110/20] via 192.168.3.1, Vlan3, 00:00:36
//OSPF 路由协议获得 192.168.3.0 网段的路由表项
C       192.168.3.0/24 is directly connected, Vlan3
C       192.168.4.0/24 is directly connected, Vlan4
Total routes are : 5 item(s)
```

实训完成后，利用"show ip route"命令，可以看到 3 台交换机的路由表项，通过直连路由 OSPF 获得了内网所有的路由表项（192.168.1.0/24、192.168.2.0/24、192.168.3.0/24 和 192.168.4.0/24）。其中类型 C 为直连路由，类型 O 为从 OSPF 路由协议获取的路由表项。

3.6 实训六　三层交换机 ACL 访问控制列表配置

任务描述

某公司中核心交换机不同端口连接了不同的接入层交换机，端口 E0/0/1 连接的接入层交换机属于公司技术部门，此端口所属网段为"192.168.1.0/24"。管理员为了限制技术部门员工的上网范围，只允许技术部门用户访问外网的 Web 服务器，决定在核心交换机上设置 ACL 来满足这一要求。网络拓扑图如图 3-6 所示。

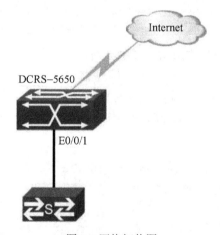

图 3-6 网络拓扑图

任务准备

ACL（Access Control List）访问控制列表是路由器和交换机接口的指令列表，用来控制端口进出的数据包。访问控制列表中包含了匹配关系、条件和查询语句。访问控制列表只是一个框架结构，其目的是为了对某种网络访问进行控制，设置完成后并没有实际意义，只有被调用之后才能显示其意义。目前有三种主要的 ACL：标准 ACL、扩展 ACL 及命名 ACL。

标准 ACL 可以阻止来自某一网络的所有通信流量，或者允许来自某一特定网络的所有通信流量。

扩展 ACL 比标准 ACL 提供了更广泛的控制范围。它基于五元组进行控制（源 IP 地址、目的 IP 地址、源端口号、目的端口号和协议号），能进行细粒度的控制。例如，可以限制某一网段的计算机只能访问某台服务器上的 WEB 服务，同时不允许访问其他服务。

命名 ACL 中使用一个字母或数字组合的字符串来代替标准 ACL 和扩展 ACL 所使用的数字。使用命名 ACL 可以用来删除某一条特定的控制条目，这样可以让我们在使用过程中方便地进行修改。在使用命名访问控制列表时不能以同一名字命名多个 ACL，不同类型的 ACL 也不能使用相同的名字，命名 ACL 创建过程中也能选择标准 ACL 或扩展 ACL 方式。神州数码交换机的防火墙应用访问控制列表时，只能应用到物理接口的入方向。

任务实施

1. 配置访问控制列表。

```
DCRS-5650-28>enable
DCRS-5650-28#config
DCRS-5650-28(config)#ip access-list extended acl    //创建命名访问控制列表 AVL
DCRS-5650-28(Config-IP-Ext-Nacl-acl)#permit udp any-source any-destination d-por
t 53                            //允许从所有源地址到所有目的地址的 DNS 请求服务
DCRS-5650-28(Config-IP-Ext-Nacl-acl)#permit tcp any-source any-destination d-por
t 80                            //允许从所有源地址到所有目的地址的 HTTP 服务
DCRS-5650-28(Config-IP-Ext-Nacl-acl)#deny ip any-source any-destination
                                //拒绝所有任何基于 IP 的通信
```

2. 在内网端口"E0/0/1"应用访问控制列表。

```
DCRS-5650-28(config)#interface e0/0/1
DCRS-5650-28(Config-If-Ethernet0/0/1)#ip access-group acl in
                //应用访问控制列表 ACL 到 Ethernet0/0/1 端口的入方向
```

3. 查看交换机配置。

```
DCRS-5650-28(config)#show running-config
no service password-encryption
hostname DCRS-5650-28
vendorlocation China
vendorContact 800-810-9119
vlan 1
ip access-list extended acl                         //访问控制列表为 ACL
  permit udp any-source any-destination d-port 53
                                //允许从所有源地址到所有目的地址的 DNS 请求服务
  permit tcp any-source any-destination d-port 80
                                //允许从所有源地址到所有目的地址的 HTTP 服务
  deny ip any-source any-destination                //拒绝所有任何基于 IP 的通信
```

```
Interface Ethernet0/0/1
  ip access-group acl in    //应用访问控制列表ACL到Ethernet0/0/1端口的入方向
Interface Ethernet0/0/2
Interface Ethernet0/0/3
Interface Ethernet0/0/4
Interface Ethernet0/0/5
Interface Ethernet0/0/6
Interface Ethernet0/0/7
Interface Ethernet0/0/8
Interface Ethernet0/0/9
Interface Ethernet0/0/10
Interface Ethernet0/0/11
Interface Ethernet0/0/12
Interface Ethernet0/0/13
Interface Ethernet0/0/14
Interface Ethernet0/0/15
Interface Ethernet0/0/16
Interface Ethernet0/0/17
Interface Ethernet0/0/18
Interface Ethernet0/0/19
Interface Ethernet0/0/20
Interface Ethernet0/0/21
Interface Ethernet0/0/22
Interface Ethernet0/0/23
Interface Ethernet0/0/24
Interface Ethernet0/0/25
Interface Ethernet0/0/26
Interface Ethernet0/0/27
Interface Ethernet0/0/28
no login
End
```

经过本实训的配置后，连接到此交换机 E0/0/1 端口的 PC 只能访问 DNS 服务和 HTTP 服务。

3.7　实训七　三层交换机 MAC 与 IP 绑定

任务描述

某公司的网络经常受到 ARP 病毒的攻击，为了保证 PC 正常上网，管理员决定在三层交换机配置 MAC 与 IP 绑定，其中 PC 连接端口 E0/0/1，PC 的 MAC 地址为"00-23-54-BF-C3-DE"，IP 地址为"192.168.1.66"。

网络拓扑图如图 3-7 所示。

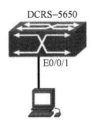

图 3-7 网络拓扑图

任务准备

AM（Access Management）即访问管理，它利用收到数据报文的信息（源 IP 地址或者源 IP+源 MAC）与配置硬件地址池（AM pool）相比较，如果找到则转发，否则丢弃。AM pool 是一个地址列表，每个地址表项对应于一个用户。每个地址表项包括了地址信息及其对应的端口。地址信息可以有两种：

- IP 地址（ip-pool），指定该端口上用户的源 IP 地址信息。
- MAC-IP 地址（mac-ip pool），指定该端口上用户的源 MAC 地址和源 IP 地址信息。

当 AM 使能的时候，AM 模块会只允许 IP 地址池内的成员源地址通过，拒绝其他所有的 IP 报文通过。交换机端口创建一个 MAC+IP 地址绑定，放到地址池中。当端口下联主机发送的 IP 报文（包含 ARP 报文）中，所含的源 IP + 源 MAC 不符合地址池中的绑定关系，此报文就将被丢弃。

这样配置的优点是配置简单，除了可以防御 ARP 攻击，还可以防御 IP 扫描等攻击，适用于信息点不多、规模不大的静态地址环境下。缺点是需要占用交换机 ACL 资源，网络管理员配置量大，终端移动性差。

任务实施

1. 启用 AM。

```
DCRS-5650-28>enable
DCRS-5650-28#config
DCRS-5650-28(config)am enable                    //全局启用 AM
```

2. 进入端口，配置 MAC-IP 绑定。

```
DCRS-5650-28(config)#interface e0/0/1
DCRS-5650-28(config-If-Ethernet0/0/1)#am port         //端口 Ethernet0/0/1
启用 AM
DCRS-5650-28(config-If-Ethernet0/0/1)#am mac-ip-pool 00-23-54-BF-C3-DE
192.168.1.66                   //在端口 Ethernet0/0/1 上绑定 MAC 和 IP 地址
DCRS-5650-28(config-If-Ethernet0/0/1)#exit
```

3. 查看交换机配置。

```
DCRS-5650-28(config)#show running-config
no service password-encryption
hostname DCRS-5650-28
vendorlocation China
vendorContact 800-810-9119
```

```
am enable                                  //全局启用 AM
vlan 1
Interface Ethernet0/0/1
 am port                                   //端口 Ethernet0/0/1 使能 AM
 am mac-ip-pool 00-23-54-bf-c3-de 192.168.1.66
                                           //在端口 Ethernet0/0/1 上绑定 MAC 和 IP 地址
Interface Ethernet0/0/2
Interface Ethernet0/0/3
Interface Ethernet0/0/4
Interface Ethernet0/0/5
Interface Ethernet0/0/6
Interface Ethernet0/0/7
Interface Ethernet0/0/8
Interface Ethernet0/0/9
Interface Ethernet0/0/10
Interface Ethernet0/0/11
Interface Ethernet0/0/12
Interface Ethernet0/0/13
Interface Ethernet0/0/14
Interface Ethernet0/0/15
Interface Ethernet0/0/16
Interface Ethernet0/0/17
Interface Ethernet0/0/18
Interface Ethernet0/0/19
Interface Ethernet0/0/20
Interface Ethernet0/0/21
Interface Ethernet0/0/22
Interface Ethernet0/0/23
Interface Ethernet0/0/24
Interface Ethernet0/0/25
Interface Ethernet0/0/26
Interface Ethernet0/0/27
Interface Ethernet0/0/28
no login
end
```

4. 查看 AM。

```
DCRS-5650-28(config)#show am
AM is enabled
Interface Ethernet0/0/1
    am port
am mac-ip-pool  00-23-54-bf-c3-de 192.168.1.66
```

交换机的 E0/0/1 端口使用 AM 后,并进行了 IP-MAC 地址的绑定,使此端口只能转发源 MAC 地址和源 IP 地址与绑定一样的数据包。

3.8 实训八 三层交换机配置 DHCP 服务器

任务描述

某公司网络管理人员经常受困于用户设置错误 IP 地址,造成不能正常上网的情况为了从这一简单任务中摆脱出来,网络管理员决定在公司内部使用 DHCP 服务器,公司网络情况如下:VLAN1 端口包含端口 1~10,IP 地址 "192.168.1.1/24",VLAN 2 端口包含端口 11~20,IP 地址为 "192.168.2.1/24",为两个 VLAN 中的 PC 分配 IP 地址、网关和 DNS。

网络拓扑图如图 3-8 所示。

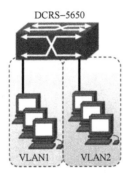

图 3-8 网络拓扑图

任务准备

DHCP（Dynamic Host Configuration Protocol）即动态主机配置协议,DHCP 用于为计算机自动提供 IP 地址、子网掩码和路由信息。网络管理员通常会分配某个范围的 IP 地址来分发给局域网上的客户机,当设备接入这个局域网时,它们会向 DHCP 服务器请求一个 IP 地址。然后 DHCP 服务器为每个请求的设备分配一个地址,直到分配该范围内的所有 IP 地址为止。已经分配的 IP 地址必须定时地延长借用期,这个延期的过程称做 "leasing",确保当客户机设备在正常地释放 IP 地址之前突然从网络断开时被分配的地址可以归还给服务器。

DHCP 的工作流程如下。

1. DHCP 发现阶段

即 DHCP 客户端查找 DHCP 服务器的阶段。客户机以广播方式（因为 DHCP 服务器的 IP 地址对于客户端来说是未知的）发送 DHCP discover 信息来查找 DHCP 服务器,即向地址 255.255.255.255 发送特定的广播信息。网络上每一台安装了 TCP/IP 的主机都会接收到这种广播信息,但只有 DHCP 服务器才会做出响应。

2. DHCP 提供阶段

即 DHCP 服务器提供 IP 地址的阶段。在网络中接收到 DHCP discover 信息的 DHCP 服务器都会做出响应,它从尚未出租的 IP 地址中挑选一个分配给 DHCP 客户端,向其发送一个包含出租的 IP 地址和其他设置的 DHCP offer 信息。

3. DHCP 选择阶段

即 DHCP 客户端选择某台 DHCP 服务器提供的 IP 地址的阶段。如果有多台 DHCP 服务器向 DHCP 客户端发送 DHCP offer 信息，则 DHCP 客户端只接受第 1 个收到的 DHCP offer 信息。然后它就以广播方式回答一个 DHCP request 信息，该信息中包含向它所选定的 DHCP 服务器请求 IP 地址的内容。之所以要以广播方式回答，是为了通知所有 DHCP 服务器，它将选择某台 DHCP 服务器所提供的 IP 地址。

4. DHCP 确认阶段

即 DHCP 服务器确认所提供的 IP 地址的阶段。当 DHCP 服务器收到 DHCP 客户端回答的 DHCP request 信息之后，它向 DHCP 客户端发送一个包含其所提供的 IP 地址和其他设置的 DHCP ACK 信息，告诉 DHCP 客户端可以使用该 IP 地址，然后 DHCP 客户端便将其 TCP/IP 与网卡绑定。另外，除 DHCP 客户端选中的服务器外，其他的 DHCP 服务器都将收回曾提供的 IP 地址。

5. DHCP 重新登录

以后 DHCP 客户端每次重新登录网络时，不需要发送 DHCP discover 信息，而是直接发送包含前一次所分配的 IP 地址的 DHCP request 信息。当 DHCP 服务器收到这一信息后，它会尝试让 DHCP 客户端继续使用原来的 IP 地址，并回答一个 DHCP ACK 信息。如果此 IP 地址已无法再分配给原来的 DHCP 客户端使用（例如此 IP 地址已分配给其他 DHCP 客户端使用），则 DHCP 服务器给 DHCP 客户端回答一个 DHCP NACK 信息。当原来的 DHCP 客户端收到此信息后，必须重新发送 DHCP discover 信息来请求新的 IP 地址。

6. DHCP 更新租约

DHCP 服务器向 DHCP 客户端出租的 IP 地址一般都有一个租借期限，期满后 DHCP 服务器便会收回该 IP 地址。如果 DHCP 客户端要延长其 IP 租约，则必须更新其 IP 租约。DHCP 客户端启动时和 IP 租约期限过半时，DHCP 客户端都会自动向 DHCP 服务器发送更新其 IP 租约的信息。

任务实施

1. 创建相应 VLAN，添加相应端口，设置相应 IP 地址。

```
DCRS-5650-28>enable
DCRS-5650-28#config
DCRS-5650-28(config)#vlan 1
DCRS-5650-28(config-Vlan1)# switchport interface e0/0/1-10
DCRS-5650-28(config-Vlan1)#exit
DCRS-5650-28(config)#vlan 2
DCRS-5650/28(Config-Vlan2)#switchport interface e0/0/11-20
DCRS-5650/28(Config-Vlan2)#exit
DCRS-5650-28(config)#interface vlan 1
DCRS-5650-28(config-if-Vlan1)#ip address 192.168.1.1 255.255.255.0
DCRS-5650-28(config-if-Vlan1)#exit
DCRS-5650-28(config)#interface vlan 2
DCRS-5650-28(config-if-Vlan2)#ip address 192.168.2.1 255.255.255.0
```

```
DCRS-5650-28(config-if-Vlan2)#exit
```

2. 配置 DHCP 服务器。

```
DCRS-5650-28(config)#service dhcp                          //启动 DHCP 服务
DCRS-5650-28(config)#ip dhcp pool vlan1                    //建立地址池 VLAN 1
DCRS-5650-28(dhcp-vlan1-config)#network-address 192.168.1.0 24
//分配网段为 192.168.1.0/24
DCRS-5650-28(dhcp-vlan1-config)#default-router 192.168.1.1
//192.168.1.0/24 网段的网关为 192.168.1.1
DCRS-5650-28(dhcp-vlan1-config)#dns-server 8.8.8.8    //DNS 服务器 IP 地址为 8.8.8.8
DCRS-5650-28(dhcp-vlan1-config)#exit
DCRS-5650-28(config)#ip dhcp pool vlan2                    //建立地址池 VLAN 2
DCRS-5650-28(dhcp-vlan2-config)#network-address 192.168.2.0 24
//分配网段为 192.168.2.0/24
DCRS-5650-28(dhcp-vlan2-config)#default-router 192.168.2.1
//192.168.1.0/24 网段的网关为 192.168.1.1
DCRS-5650-28(dhcp-vlan2-config)#dns-server 8.8.8.8
//DNS 服务器 IP 地址为 8.8.8.8
DCRS-5650-28(dhcp-vlan2-config)#exit
```

3. 查看交换机配置。

```
DCRS-5650-28(config)#show running-config
no service password-encryption
hostname DCRS-5650-28
vendorlocation China
vendorContact 800-810-9119
service dhcp                                    // DHCP 服务已经启动
ip dhcp pool vlan2                              //地址池 VLAN 2
 network-address 192.168.2.0 255.255.255.0      //分配网段为 192.168.2.0/24
 default-router 192.168.2.1                     //网关 IP 地址为 192.168.2.1
 dns-server 8.8.8.8                             //DNS 服务器 IP 地址为 8.8.8.8
ip dhcp pool vlan1                              //地址池 VLAN 1
 network-address 192.168.1.0 255.255.255.0      //分配网段为 192.168.1.0/24
 default-router 192.168.1.1                     //网关 IP 地址为 192.168.1.1
 dns-server 8.8.8.8                             //DNS 服务器 IP 地址为 8.8.8.8
vlan 1
vlan 2
Interface Ethernet0/0/1
Interface Ethernet0/0/2
略
Interface Ethernet0/0/9
Interface Ethernet0/0/10
Interface Ethernet0/0/11
 switchport access vlan 2
Interface Ethernet0/0/12
 switchport access vlan 2
```

```
Interface Ethernet0/0/13
 switchport access vlan 2
Interface Ethernet0/0/14
 switchport access vlan 2
Interface Ethernet0/0/15
 switchport access vlan 2
Interface Ethernet0/0/16
 switchport access vlan 2
Interface Ethernet0/0/17
 switchport access vlan 2
Interface Ethernet0/0/18
 switchport access vlan 2
Interface Ethernet0/0/19
 switchport access vlan 2
Interface Ethernet0/0/20
 switchport access vlan 2
Interface Ethernet0/0/21
Interface Ethernet0/0/22
略
Interface Ethernet0/0/27
Interface Ethernet0/0/28
interface Vlan1
 ip address 192.168.1.1 255.255.255.0
interface Vlan2
 ip address 192.168.2.1 255.255.255.0
no login
end
```

通过以上对交换机的配置后，交换机就能为VLAN1和VLAN2中的PC分配IP地址、网关和DNS服务器，同时要求客户端的本地连接设置为自动获取IP地址。

3.9　实训九　三层交换机DHCP中继配置

任务描述

某公司网络中配置了一台三层交换机（名称为DHCP服务器）作为专用的DHCP服务器，另外一台三层交换机（名称为DHCP中继代理）通过端口E0/0/24与DHCP服务器的端口E0/0/24相连。DHCP服务器中VLAN1的IP地址为"192.168.1.1/24"。DHCP中继代理有两个VLAN，其中VLAN1的IP地址为"192.168.1.2/24"，VLAN2的IP地址为"192.168.5.1/24"。两台三层交换机之间以启用RIP协议获取路由表项。网络管理员在DHCP服务器上设置地址池为"192.168.5.0/24"，在DHCP中继代理的VLAN2接口上启用DHCP中继，来帮助VLAN2中的PC获取IP地址。

网络拓扑图如图 3-9 所示。

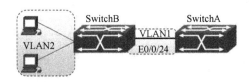

图 3-9 网络拓扑图

任务准备

DHCP Relay Agent：DHCP Relay Agent 即 DHCP 中继代理。如果 DHCP 客户机与 DHCP 服务器在同一个物理网段，则客户机可以正确地获得动态分配的 IP 地址。如果不在同一个物理网段，则需要 DHCP 中继代理。用 DHCP 中继代理可以传递消息到不在同一个物理子网的 DHCP 服务器，也可以将服务器的消息传回给不在同一个物理子网的 DHCP 客户机，帮助客户机正确地获得 IP 地址。

任务实施

SwitchA

1. 创建相应 VLAN，设置相应 IP 地址。

```
DCRS-5650-28>enable
DCRS-5650-28#config
DCRS-5650-28(config)#hostname SwitchA
SwitchA(config)#vlan 1
SwitchA(Config-Vlan1)#exit
SwitchA(config)#interface vlan 1
SwitchA(Config-if-Vlan1)#ip address 192.168.1.1 255.255.255.0
SwitchA(Config-if-Vlan1)#exit
```

2. 配置 RIP 协议

```
SwitchA(config)#router rip                              //进入RIP视图
SwitchA(config-router)#version 2                        //设置为RIPv2
SwitchA(config-router)#network 192.168.1.0/24
//在192.168.1.0网段启用RIP路由协议
```

3. 配置 DHCP 服务器。

```
SwitchA(config)#service dhcp
SwitchA(config)#ip dhcp pool 123
SwitchA(dhcp-123-config)#network-address 192.168.5.0 24
SwitchA(dhcp-123-config)#default-router 192.168.5.1
SwitchA(dhcp-123-config)#dns-server 8.8.8.8
SwitchA(dhcp-123-config)#exit
```

4. 查看交换机配置。

```
SwitchA(config)#show running-config
no service password-encryption
```

```
hostname SwitchA
vendorlocation China
vendorContact 800-810-9119
service dhcp
ip dhcp pool 123
 network-address 192.168.5.0 255.255.255.0
 default-router 192.168.5.1
 dns-server 8.8.8.8
vlan 1
Interface Ethernet0/0/1
Interface Ethernet0/0/2
Interface Ethernet0/0/3
Interface Ethernet0/0/4
Interface Ethernet0/0/5
Interface Ethernet0/0/6
Interface Ethernet0/0/7
Interface Ethernet0/0/8
Interface Ethernet0/0/9
Interface Ethernet0/0/10
Interface Ethernet0/0/11
Interface Ethernet0/0/12
Interface Ethernet0/0/13
Interface Ethernet0/0/14
Interface Ethernet0/0/15
Interface Ethernet0/0/16
Interface Ethernet0/0/17
Interface Ethernet0/0/18
Interface Ethernet0/0/19
Interface Ethernet0/0/20
Interface Ethernet0/0/21
Interface Ethernet0/0/22
Interface Ethernet0/0/23
Interface Ethernet0/0/24
Interface Ethernet0/0/25
Interface Ethernet0/0/26
Interface Ethernet0/0/27
Interface Ethernet0/0/28
interface Vlan1
 ip address 192.168.1.1 255.255.255.0
router rip
 network 192.168.1.0/24
no login
end
```

SwitchB

1. 创建相应 VLAN，添加相应端口，设置相应 IP 地址。

```
DCRS-5650-28>enable
DCRS-5650-28#config
```

```
DCRS-5650-28#hostname SwitchB
SwitchB(config)#vlan 1
SwitchB (Config-Vlan1)#switchport interface e0/0/24
SwitchB (Config-Vlan1)#exit
SwitchB (config)#vlan 2
SwitchB (Config-Vlan2)#switchport interface ethernet 0/0/1-10
SwitchB (Config-Vlan2)#exit
SwitchB (config)#interface vlan 1
SwitchB (Config-if-Vlan1)#ip address 192.168.1.2 255.255.255.0
SwitchB (Config-if-Vlan1)#exit
SwitchB (config)#interface vlan 2
SwitchB (Config-if-Vlan2)#ip address 192.168.5.1 255.255.255.0
SwitchB (Config-if-Vlan2)#exit
```

2. 配置 RIP 协议

```
SwitchB(config)#router rip                              //进入 RIP 视图
SwitchB(config-router)#version 2                        //设置为 RIPv2
SwitchB(config-router)#network 192.168.1.0/24
//在 192.168.1.0 网段启用 RIP 路由协议
SwitchB(config-router)#network 192.168.5.0/24
//在 192.168.5.0 网段启用 RIP 路由协议
```

3. 配置 DHCP 中继代理。

```
SwitchB (config)#service dhcp                           //启动 DHCP 服务
SwitchB (config)#ip forward-protocol udp bootps  //转发 DHCP 数据包
SwitchB (config)#interface vlan 2
SwitchB (Config-if-Vlan5)#ip helper-address 192.168.1.1
//在交换机的 VLAN2 上启用 DHCP 中继代理,转发到 192.168.1.1
SwitchB (Config-if-Vlan5)#exit
```

4. 查看交换机配置。

```
 SwitchB (config)#show running-config
no service password-encryption
hostname SwitchB
ip forward-protocol udp 67
service dhcp
vlan 1
vlan 2
Interface Ethernet0/0/1
 switchport access vlan 2
Interface Ethernet0/0/2
 switchport access vlan 2
Interface Ethernet0/0/3
 switchport access vlan 2
Interface Ethernet0/0/4
 switchport access vlan 2
Interface Ethernet0/0/5
 switchport access vlan 2
```

```
Interface Ethernet0/0/6
 switchport access vlan 2
Interface Ethernet0/0/7
 switchport access vlan 2
Interface Ethernet0/0/8
 switchport access vlan 2
Interface Ethernet0/0/9
 switchport access vlan 2
Interface Ethernet0/0/10
 switchport access vlan 2
Interface Ethernet0/0/11
Interface Ethernet0/0/12
Interface Ethernet0/0/13
Interface Ethernet0/0/14
Interface Ethernet0/0/15
Interface Ethernet0/0/16
Interface Ethernet0/0/17
Interface Ethernet0/0/18
Interface Ethernet0/0/19
Interface Ethernet0/0/20
Interface Ethernet0/0/21
Interface Ethernet0/0/22
Interface Ethernet0/0/23
Interface Ethernet0/0/24
Interface Ethernet0/0/25
Interface Ethernet0/0/26
Interface Ethernet0/0/27
Interface Ethernet0/0/28
interface Vlan1
 ip address 192.168.1.2 255.255.255.0
interface Vlan2
 ip address 192.168.5.1 255.255.255.0
  !forward protocol udp 67(active)!          //启动 DHCP 中继代理
  ip helper-address 192.168.1.1              //设置 DHCP 服务器为 192.168.1.1
router rip
 network 192.168.1.0/24
 network 192.168.5.0/24
no login
end
```

经过对这两个交换机的配置后，DHCP 中继代理的 VLAN2 中的 PC 就能获得 DHCP 服务器分配的 IP 地址，但是反应速度没有直接连接 DHCP 服务器快。

注意：DHCP 服务器要为不相连网段提供服务，它的路由表中一定要有到那个网段的路由表项。

项目 4

路由器的管理和维护

教学目标

通过本章的学习,学生可以了解路由器的基础配置方式、方法和基本配置命令,为接下来的学习路由器的具体应用配置打好基础。

能力目标

了解路由器在网络中的作用
熟悉路由器的基础配置
掌握路由器的基础配置命令

知识目标

熟悉路由器的端口分类
熟悉路由器的配置模式
熟悉路由器的配置方法

主要教学内容

路由器的管理
路由器的物理和逻辑端口
路由器配置文件的相关操作

4.1 实训一 认识路由器模块和端口

任务 认识 DCR-2626 路由器的模块和端口。

任务描述

前面我们学习了交换机的相关知识和配置。从本章开始,我们将进入路由器部分。首先,来学习路由器的模块和端口,对 DCR-2626 模块化路由器有一个总体的认识。

任务准备

实验所需设备为 DCR-2626 路由器一台。

任务实施

DCR-2626 模块化路由器的标配端口由 5 个部分组成:2 个单口快速以太网端口,1 个 Console 端口,1 个 AUX 端口,1 个双串口(支持 V.28/V.35)。详细说明见表 4-1。

表4-1 标配端口表

标配端口名称	特　点
单口快速以太网端口	速率 10/100M 自适应,UTP (RJ45) 接口,带 ACT、Link、100Mbps 指示灯
Console 端口	速率 1200~115200bps,RJ45 接口,无指示灯
AUX 端口	速率 1200~57600bps,RJ45 接口,无指示灯
高速串口	速率 2400~2048000bps,DB60 接口。

1. DCR-2626 前面板如图 4-1 所示。

图4-1 路由器面板

DCR-2626 前面板有 4 个 SLOT 模块,说明见表 4-2。

表4-2 面板模块介绍

模块编号	英文名称	中文名称	说明
1	SLOT1	1 号扩展插槽	支持单串口卡、单以太网卡、双串口卡、双以太网卡、单以太网+单串口复合卡、两路 E&M、FXS、FXO 语音卡、八路异步卡、单路 ISDN BRI S/T 接口卡、单路 E1 卡、DTU 接口卡、一路 ISDN U 接口卡、一路 MODEM 卡
2	SLOT2	2 号扩展插槽	支持两路 E&M、FXS、FXO 语音卡、八路异步卡
3	SLOT3	3 号扩展插槽	支持两路 E&M、FXS、FXO 语音卡、八路异步卡
4	SLOT4	4 号扩展插槽	支持两路 E&M、FXS、FXO 语音卡、八路异步卡

2. DCR-2626 后面板如图 4-2 所示。

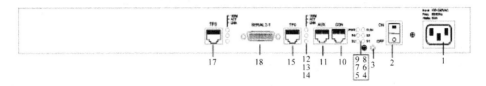

图4-2 路由器后面板

相应编号对应设备的具体说明见表 4-3。

表4-3 路由器后面板端口

编 号	英 文 名 称	中 文 名 称	说 明
1		交流电源插座	AC 100～240V
2	Power	电源开关	向上按为开，向下按为关
3		接地柱	需良好接地
4	S1	槽位 1 状态	亮表示槽位 1 有扩展模块；灭表示槽位 1 没插任何模块
5	S2	槽位 2 状态	亮表示槽位 2 有扩展模块；灭表示槽位 2 没插任何模块
6	S3	槽位 3 状态	亮表示槽位 3 有扩展模块；灭表示槽位 3 没插任何模块
7	S4	槽位 4 状态	亮表示槽位 4 有扩展模块；灭表示槽位 4 没插任何模块
8	RUN	系统运行灯	系统正常启动后，该灯闪烁
9	PWR	系统电源灯	系统电源打开后，该灯点亮
10	CON	Console 端口	连接配置线缆
11	AUX	AUX 端口	备份使用
12	100M	100M 以太网指示灯	当 10/100M 以太网口工作于 100M 方式时该灯会亮
13	LINK	10/100M 以太网口连接有效指示灯	当以太网经双绞线与 Hub（集线器）有效连接后，该灯会由灭转亮
14	ACT	10/100M 以太网口收/发数据指示灯	当以太网口有数据收/发时，该灯会闪烁
15	TP0	10/100M 以太网双绞线端口	通过该端口与局域网（以太网）以双绞线方式连接
16	SERIAL2-1	高速串口	速率 2400～2048000bps，DB60 接口
17	TP3	快速以太网交换接口	通过该端口与局域网（以太网）以双绞线方式连接

4.2 实训二 路由器带外及带内管理

任务 了解路由器的管理方式，掌握带内、带外管理方式。

任务描述

认识了路由器的模块和端口，接下来我们来学习如何登录管理路由器。

任务准备

实验所需设备为 DCS-2626 路由器一台，PC 一台，配置线缆一根，直通线缆一根。实验拓扑图如图 4-3 所示。

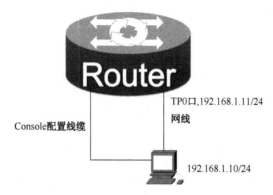

图4-3 实验拓扑图

任务实施

1. 带外管理。路由器的带外管理也是通过"超级终端程序"，本书前面已经介绍过，在此不再赘述。配置好超级终端，登录路由器，如图 4-4 所示。

图4-4 路由器登录界面

我们也可以使用"show version"命令来查看该路由器的版本信息。但请注意，该命令在特权模式下才能使用，如图 4-5 所示。

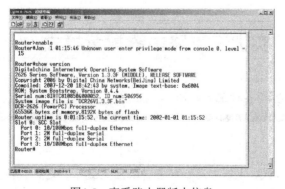

图4-5 查看路由器版本信息

2. 带内管理。和交换机的带内管理一样,路由器的带内管理也是通过 Telnet 命令来实现的。

```
Router>enable                                           //进入特权模式
Router#Jan  1 00:10:30 Unknown user enter privilege mode from console 0,
level =15
Router#config                                           //进入全局配置模式
Router_config#interface fastEthernet 0/0        //进入 TP0 端口模式
Router_config_f0/0#ip address 192.168.1.11 255.255.255.0  //设置 IP 地址
Router_config#ping 192.168.1.10                 //测试 PC 机连通情况
PING 192.168.1.10 (192.168.1.10): 56 data bytes
!!!!!                                           //5个!号,ping 通
Router_config#username test privilege 15 password 0 test      // 为
路由器设置授权 Telnet 用户设置用户名"test",配置优先级 15(0_15,15 的优先级最高),配置明文密码为"test"(0 是明文密码,7 是加密密码并且密码长度必须为 32 位)
Router_config#aaa authentication login default local   //设置验证方式为本地
```

配置主机的 IP 地址 192.168.1.10,在本实验中要与路由器的 IP 地址在一个网段。验证方法是使用 Telnet 登录。输入正确的用户名和密码后,登录成功,如图 4-6 所示。

图4-6 验证界面

路由器提示:

```
Router_config#aaa authentication lJan  1 01:23:27 User test logged in from
192.168.1.10 onvty0
```

登录成功,实现了通过 Telnet 服务实现带内管理。

4.3 实训三 路由器配置模式

任 务 掌握路由器的各种配置视图

任务描述

学习了带内管理方式,大家对路由的各种配置视图有了更多了解,本实训来学习路由器的各种配置视图。

任务准备

实验所需设备为 DCS-2626 路由器一台，PC 一台，配置线缆一根。实验拓扑图如图 4-7 所示。

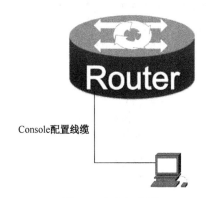

图4-7 实验拓扑图

任务实施

1. 一般用户视图

当我们使用超级终端登录路由器，首先进入的就是一般用户视图。它的提示符为"＞"，可用的命令如图 4-8 所示，命令较少。

图4-8 一般用户视图

2. 特权模式视图

当我们在一般用户视图中，输入"enable"命令就可以进入特权视图，在这里可以把语言模式切换为中文，查看当前所有配置。提示符为"#"，可用命令如图 4-9 所示。

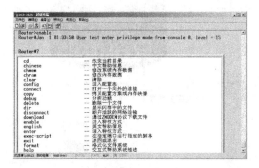

图4-9 特权模式视图

3. 全局配置模式

在特权用户视图输入"config"命令就会进入全局配置视图,提示符为"Router_config#",在这里我们可以完成所有对路由器的配置。其常用命令如图 4-10 所示。

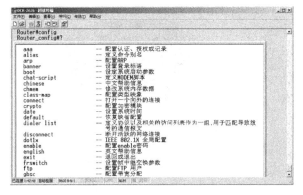

图4-10 全局配置模式

以上就是路由器常见的三种配置模式,它们的退出命令都是"exit"。

4.4 实训四 管理路由器账号

任务 为路由器设置登录账号。

任务描述

和交换机一样,在日常工作中,为了避免有人登录路由器并修改配置文件,网络管理员一般都需要为路由器设置管理密码。

任务准备

实验所需设备为 DCS-2626 路由器一台,PC 一台,配置线缆一根。实验拓扑图如图 4-11 所示。

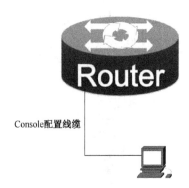

图4-11 实验拓扑图

任 务 实 施

1. 在全局配置模式下设置特权用户口令。

```
Router>enable                                      //进入特权配置模式
Router#Jan  1 01:59:31 User test enter privilege mode from console 0, level
= 15
Router#config                                      //进入全局配置模式
Router_config#enable password 0 admin              //设置特权配置模式密码为admin
Router_config #write                               //保存配置
```

2. 验证实验。

```
Router_config #exit                                //退出特权用户配置模式
Router_config >enable                              //进入特权模式
Password:                                          //提示输入登录密码
Router_config #                                    //密码验证通过，进入特权模式
```

4.5 实训五 管理路由器系统文件

任 务 备份路由器的系统文件。

任 务 描 述

当我们完成对路由器的配置以后，接下来的工作就是把配置文件和系统文件从路由器里复制出来另外存放，防止意外情况的发生。当路由器出现故障以后，我们可以直接把备份的文件下载到路由器上，让网络能够继续正常运作。TFTP 相对于 FTP 的优点是提供简单、开销不大的文件传输服务，更加适合路由器备份文件使用。我们使用 tftpd32 作为 TFTP 服务器。

任 务 准 备

实验所需设备为 DCR-2626 路由器一台，PC 一台，配置线缆一根，直通线缆一根。实验拓扑图如图 4-12 所示。

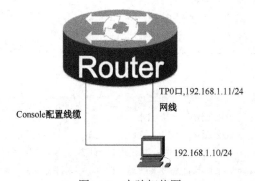

图4-12　实验拓扑图

任务实施

实验步骤：

1. 配置 TFTP 服务器

双击 tftpd32.exe，出现 TFTP 服务器的主界面如图 4-13 所示。

图4-13 TFTP服务器

在主界面中，我们看到该服务器的根目录是 "D:\ DCR-2626 备份"，服务器的 IP 地址也自动出现在第二行：192.168.1.10。可以点击 "Browse" 按钮更改当前目录，单击 "确定" 按钮进行保存确认。也可以单击 "Settings" 按钮进行其他设置。到此 TFTP 服务器搭建好了，可以将它最小化到右下角的工具栏中。

2. 给路由器设置 IP 地址即管理 IP。验证主机与路由器是否连通。

3. 查看需要备份的文件。

```
Router#dir
Directory of /:
0    DCR26V1.3.3F.bin      <FILE>    5054463   Tue Jan  1 00:25:00 2002   //系统文件
1    startup-config        <FILE>         669   Tue Jan  1 01:45:41 2002   //配置文件
```

4. 备份配置文件。

```
Router#copy startup-config tftp:                    //复制 tartup-config 文件
Remote-server ip address[]?192.168.1.10             //输入 tftp 服务器 IP 地址
Destination file name[startup-config]?              //给备份文件命名，默认不变
#
TFTP:successfully send 2 blocks ,669 bytes          //备份成功
```

5. 备份系统文件。

```
Router#copy flash: tftp                             //上传到 tftp 服务器
Source file name[]?DCR26V1.3.3F.bin                 //输入系统文件名
Remote-server ip address[]?192.168.1.10             //输入 tftp 服务器 IP 地址
Destination file name[DCR26V1.3.3F.bin]?            //给备份文件命名，默认不变
###############################################################
#
```

```
TFTP:successfully send 9872 blocks ,5054463 bytes          //上传成功
```

6. 验证实验。

去"D:\ DCR-2626 备份"目录查看上传的文件和大小，如图 4-14 所示。

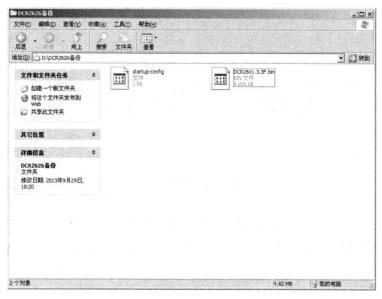

图4-14　查看备份文件

注意观察两个文件的大小，与用路由器"show flash"命令查看到的文件一致。

项目 5

路由器的基础应用

教学目标

通过本章的学习,学生可以了解路由器在网络中的具体应用的配置命令以及针对相应配置的方式和方法。掌握各种路由协议在路由器中的配置。

能力目标

了解路由器的常见应用
熟悉路由器常见应用的配置方法
掌握路由器常见应用的配置命令

知识目标

熟悉路由协议的应用环境
熟悉路由器的配置方式和配置思路
熟悉路由器各项应用的配置步骤

主要教学内容

路由协议的配置
端口封装的配置
特殊类型路由的配置

5.1 实训一 路由器单臂路由配置

任务描述

某公司中二层交换机上设置了 3 个 VLAN，分别为 VLAN100、VLAN200 和 VLAN300，端口 E0/0/1～5 属于 VLAN100，端口 E0/0/6～10 属于 VLAN200，端口 E0/0/11～15 属于 VLAN300。虽然隔绝了这 3 个 VLAN 的广播域，但同时造成了这 3 个 VLAN 内的 PC 不能互相通信。为了解决这一问题，管理员添加了一台路由器，连接路由器 F0/0 端口和交换机 E0/0/24 端口，设置交换机 E0/0/24 端口为 Trunk 端口，设置路由器为单臂路由，配置路由器 F0/0 端口的 3 个子端口的 IP 地址和封装 VLAN 号见表 5-1。

表 5-1 VLAN 参数表

路由器端口号	封装 VLAN 号	IP 地址
F0/0.1	dot1Q100	192.168.100.1/24
F0/0.2	dot1Q200	192.168.200.1/24
F0/0.3	dot1Q300	192.168.300.1/24

网络拓扑图如图 5-1 所示。

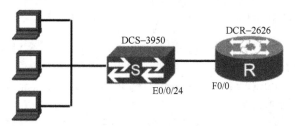

图 5-1 网络拓扑图

任务准备

单臂路由：单臂路由（router-on-a-stick）是指在路由器的一个端口上通过配置子端口（或"逻辑端口"）的方式，实现原来相互隔离的不同 VLAN（虚拟局域网）之间的互联互通。其缺点是容易成为网络单点故障，配置复杂，现实使用意义有限。

任务实施

1. 配置交换机，将相应端口加入到相应 VLAN，并把 E0/0/24 端口设置为"Trunk"端口。（略）

2. 进入端口 F0/0.1，并进行如下配置。

```
Router>enable
Router#config
Router_config#interface f0/0.1                              //进入F0/0.1子端口
Router_config_f0/0.1#ip address 192.168.100.1 255.255.255.0
//设置IP地址
```

```
Router_config_f0/0.1#encapsulation dot1Q 100         //封装VLAN号
Router_config_f0/0.1#exit
```

3. 进入端口F0/0.2，并进行如下配置。

```
Router_config#interface f0/0.2                       //进入F0/0.2子端口
Router_config_f0/0.2#ip address 192.168.200.1 255.255.255.0
  //设置IP地址
Router_config_f0/0.2#encapsulation dot1Q 200         //封装VLAN号
Router_config_f0/0.2#exit
```

4. 进入端口F0/0.3，并进行如下配置。

```
Router_config#interface f0/0.3                       //进入F0/0.3子端口
Router_config_f0/0.3#ip address 192.168.300.1 255.255.255.0//设置IP地址
Router_config_f0/0.3#encapsulation dot1Q 300         //封装VLAN号
Router_config_f0/0.3#exit
```

5. 查看路由表。

```
Router_config#show ip route                          //显示路由表
Codes: C - connected, S - static, R - RIP, B - BGP, BC - BGP connected
       D - BEIGRP, DEX - external BEIGRP, O - OSPF, OIA - OSPF inter area
       ON1 - OSPF NSSA external type 1, ON2 - OSPF NSSA external type 2
       OE1 - OSPF external type 1, OE2 - OSPF external type 2
       DHCP - DHCP type, L1 - IS-IS level-1, L2 - IS-IS level-2
VRF ID: 0
C    192.168.100.0/24     is directly connected, FastEthernet0/0.1
//192.168.100.0网段路由
C    192.168.200.0/24     is directly connected, FastEthernet0/0.2
//192.168.200.0网段路由
C    192.168.300.0/24     is directly connected, FastEthernet0/0.3
//192.168.300.0网段路由
```

6. 查看路由器配置。

```
Router_config#show running-config
!version 1.3.3H
service timestamps log date
service timestamps debug date
no service password-encryption
gbsc group default
interface FastEthernet0/0
 no ip address
 no ip directed-broadcast
interface FastEthernet0/0.1                          //路由器端口FastEthernet0/0.1
 ip address 192.168.100.1 255.255.255.0  //端口FastEthernet0/0.1的IP地址
 no ip directed-broadcast
 encapsulation dot1Q 100                             //封装VLAN号为100
 bandwidth 100000
 delay 1
```

```
 interface FastEthernet0/0.2                //路由器端口FastEthernet0/0.2
  ip address 192.168.200.1 255.255.255.0    //端口FastEthernet0/0.2的IP地址
  no ip directed-broadcast
  encapsulation dot1Q 200                   //封装VLAN号为200
  bandwidth 100000
  delay 1
 interface FastEthernet0/0.3                //路由器端口FastEthernet0/0.3
  ip address 192.168.300.1 255.255.255.0    //端口FastEthernet0/0.3的IP地址
  n2o ip directed-broadcast
  encapsulation dot1Q 300                   //封装VLAN号为300
  bandwidth 100000
  delay 1
 interface FastEthernet0/3
  no ip address
 no ip directed-broadcast
 interface Serial0/1
  no ip address
  no ip directed-broadcast
 interface Serial0/2
  no ip address
  n2o ip directed-broadcast
 interface Async0/0
  no ip address
  no ip directed-broadcast
```

7. 客户机可以通过 ping 其他 VLAN 的 PC，测试网络的联通性。

本实训中路由器的 FastEthernet0/0 端口，通过在其子端口上设置 IP 和 VLAN ID 号，拥有了同交换机 Trunk 端口一样的功能，并利用自身的路由功能，实现了不同 VLAN 间的路由。

注意：在本实训中客户机的本地连接的网关设置一定要正确。

5.2 实训二 路由器静态路由配置

任务描述

某公司有 3 台路由器，按如图 5-2 所示拓扑结构进行连接，网络中有多个 IP 地址段，管理员在所有路由器上配置静态路由来保证各个网段之间所有 PC 能互联互通。路由器的端口 IP 地址见表 5-2。

表 5-2 各路由器的端口 IP 地址

Router-A		Router-B		Router-C	
端口	IP 地址	端口	IP 地址	端口	IP 地址
F0/0	192.168.2.1	F0/0	192.168.3.1	F0/0	192.168.2.2
F0/3	192.168.1.1	F0/3	192.168.1.2	F0/3	192.168.4.1
S0/1 DTE	211.159.127.32				

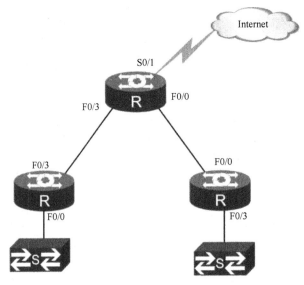

图 5-2　网络拓扑图

任务准备

静态路由：静态路由是指由网络管理员手工配置的路由信息。当网络的拓扑结构或链路的状态发生变化时，网络管理员需要手工去修改路由表中相关的静态路由信息。静态路由信息在默认情况下不会传递给其他的路由器。

默认路由：在路由表中，默认路由以目的网络为 0.0.0.0、子网掩码为 0.0.0.0 的形式出现。默认路由是一条特殊的静态路由，就是在没有找到匹配的路由时使用的路由。如果数据包的目的地址不能与任何路由相匹配，那么系统将使用默认路由转发该数据包。

任务实施

Router-A

1. 设置路由器 Router-A 各端口 IP 地址。

```
Router>enable
Router#config
Router_config#hostname Router-A                    //设置路由器名称为Router-A
Router-A_config#interface f0/0                     //进入端口F0/0
Router-A_config_f0/0#ip address 192.168.2.1 255.255.255.0  //配置端口IP地址
Router-A_config_f0/0#exit
R2outer-A_config#interface f0/3                    //进入端口F0/3
Router-A_config_f0/3#ip address 192.168.1.1 255.255.255.0  //配置端口IP地址
Router-A_config_f0/3#exit
Router-A_config#interface s0/1                     //进入端口S0/1
Router-A_config_s0/1#ip address 211.159.127.32 255.255.255.0
                                                   //配置端口IP地址
Router-A_config_s0/1#exit
```

2. 设置路由器 Router-A 上静态路由。

```
Router-A_config#ip route 0.0.0.0 0.0.0.0 211.159.127.32
                //配置默认路由指向211.159.127.32
Router-A_config#ip route 192.168.3.0 255.255.255.0 192.168.1.2
                //配置到192.168.3.0网段静态路由，下一跳为192.168.1.2
Router-A_config#ip route 192.168.4.0 255.255.255.0 192.168.2.2
                //配置到192.168.4.0网段静态路由，下一跳为192.168.2.2
```

3. 查看路由器 Router-A 的路由表。

```
Router-A_config#show ip route
Codes: C - connected, S - static, R - RIP, B - BGP, BC - BGP connected
       D - DEIGRP, DEX - external DEIGRP, O - OSPF, OIA - OSPF inter area
       ON1 - OSPF NSSA external type 1, ON2 - OSPF NSSA external type 2
       OE1 - OSPF external type 1, OE2 - OSPF external type 2
       DHCP - DHCP type
VRF ID: 0
S       0.0.0.0/0           [1,0] via 211.159.127.32(on Serial0/1)
                                    //默认路由
C       192.168.1.0/24      is directly connected, FastEthernet0/3
C       192.168.2.0/24      is directly connected, FastEthernet0/0
S       192.168.3.0/24      [1,0] via 192.168.1.2(on FastEthernet0/3)
                                    //192.168.3.0网段静态路由
S       192.168.4.0/24      [1,0] via 192.168.2.2(on FastEthernet0/0)
                                    // 192.168.4.0网段静态路由
C       211.159.127.0/24    is directly connected, Serial0/1
```

4. 查看路由器配置。

```
Router-A_config#show running-config              //显示路由器当前配置文件
!version 1.3.3F
service timestamps log date
service timestamps debug date
no service password-encryption
hostname Router-A
gbsc group default
interface FastEthernet0/0
 ip address 192.168.2.1 255.255.255.0
 no ip directed-broadcast
interface FastEthernet0/3
 ip address 192.168.1.1 255.255.255.0
 no ip directed-broadcast
interface Serial0/1
 ip address 211.159.127.32 255.255.255.0
 no ip directed-broadcast
interface Serial0/2
 no ip address
 no ip directed-broadcast
interface Async0/0
```

```
 no ip address
 no ip directed-broadcast
ip route default 211.159.127.32                        //默认路由
ip route 192.168.3.0 255.255.255.0 192.168.1.2  //192.168.3.0网段静态路由
ip route 192.168.4.0 255.255.255.0 192.168.2.2  //192.168.4.0网段静态路由
```

Router-B

1. 如表 5-2 所示，设置 IP 地址。

```
Router>enable
Router#config
Router_config#hostname Router-B                    //设置路由器名称为Router-B
Router-B_config#interface f0/0
Router-B_config_f0/0#ip address 192.168.3.1 255.255.255.0  //设置IP地址
Router-B_config_f0/0#exit
Router-B_config#interface f0/3
Router-B_config_f0/3#ip address 192.168.1.2 255.255.255.0  //设置IP地址
Router-B_config_f0/3#exit
```

2. 设置静态路由。

```
Router-B_config#ip route 0.0.0.0 0.0.0.0 192.168.1.1
                                   //设置默认路由指向192.168.1.1
```

3. 查看路由表。

```
Router-B_config#show ip route                          //显示路由表
Codes: C - connected, S - static, R - RIP, B - BGP, BC - BGP connected
       D - BEIGRP, DEX - external BEIGRP, O - OSPF, OIA - OSPF inter area
       ON1 - OSPF NSSA external type 1, ON2 - OSPF NSSA external type 2
       OE1 - OSPF external type 1, OE2 - OSPF external type 2
       DHCP2 - DHCP type, L1 - IS-IS level-1, L2 - IS-IS level-2
VRF ID: 0
S       0.0.0.0/0         [1,0] via 192.168.1.1(on FastEthernet0/3)
                                   //默认路由指向192.168.1.1
C       192.168.1.0/24    is directly connected, FastEthernet0/3
C       192.168.3.0/24    is directly connected, FastEthernet0/0
```

4. 查看路由器配置。

```
Router-B_config#show running-config                //显示路由器当前配置文件
!version 1.3.3H
service timestamps log date
service timestamps debug date
no service password-encryption
hostname Router-B
gbsc group default
interface FastEthernet0/0
 ip address 192.168.3.1 255.255.255.0
 no ip directed-broadcast
```

```
interface FastEthernet0/3
 ip address 192.168.1.2 255.255.255.0
 no ip directed-broadcast
interface Serial0/1
 no ip address
 no ip directed-broadcast
interface Serial0/2
 no ip address
 no ip directed-broadcast
 physical-layer speed 2048000
interface Async0/0
 no ip address
 no ip directed-broadcast
ip route default 192.168.1.1                    //默认路由
```

Router-C

1. 设置路由器 Router-C 的 IP 地址。

```
Router>enable
Router#config
Router_config#hostname Router-C                 //设置路由器名称为Router-C
R2outer-C_config#interface f0/0
Router-C_config_f0/0#ip address 192.168.2.2 255.255.255.0    //设置IP地址
Router-C_config_f0/0#exit
Router-C_config#interface f0/3
Router-C_config_f0/3#ip address 192.168.4.1 255.255.255.0
  //设置IP地址
```

2. 设置静态路由。

```
Router-B_config#ip route 0.0.0.0 0.0.0.0 192.168.2.1
                              //设置默认路由指向192.168.2.1
```

3. 查看路由表。

```
Router-C_config#show ip route                   //显示路由表
Codes: C - connected, S - static, R - RIP, B - BGP, BC - BGP connected
       D - BEIGRP, DEX - external BEIGRP, O - OSPF, OIA - OSPF inter area
       ON1 - OSPF NSSA external type 1, ON2 - OSPF NSSA external type 2
       OE1 - OSPF external type 1, OE2 - OSPF external type 2
       DHCP - DHCP type, L1 - IS-IS level-1, L2 - IS-IS level-2
VRF ID: 0
S       0.0.0.0/0        [1,0] via 192.168.2.1(on FastEthernet0/0)
C       192.168.2.0/24    is directly connected, FastEthernet0/0
C       192.168.4.0/24    is directly connected, FastEthernet0/3
```

4. 查看路由器配置。

```
Router-C_config#show running-config             //显示路由器当前配置文件
!version 1.3.3H
service timestamps log date
```

```
service timestamps debug date
no service password-encryption
hostname Router-C
gbsc group default
interface FastEthernet0/0
 ip address 192.168.2.2 255.255.255.0
 no ip directed-broadcast
interface FastEthernet0/3
 ip address 192.168.4.1 255.255.255.0
 no ip directed-broadcast
interface Serial0/1
 no ip address
 no ip directed-broadcast
interface Serial0/2
 no ip address
 no ip directed-broadcast
interface Async0/0
 no ip address
 no ip directed-broadcast
ip route default 192.168.2.1                    //默认路由
```

经过上述对 3 个路由器的配置，每个路由器的路由表中都包含了内网所有网段，或有默认路由，内网所有用户之间实现了互联互通，可以通过 ping 测试。

注意：本实训中 Router-B 和 Router-C 两台路由器中仅设置了一条默认路由，这种设置必须在本路由器是末节路由器的情况下才能使用。

5.3 实训三 路由器 RIP 协议配置

任务描述

某公司有 3 台路由器，按如图 5-3 所示拓扑结构进行连接，网络中有多个 IP 地址段。管理员发现在网络的拓扑结构或链路的状态发生变化时，需要手工去修改路由表中相关的静态路由信息，于是决定在所有路由器配置动态路由协议 RIP 来保证各个网段之间所有 PC 能互联互通。路由器的端口 IP 地址见表 5-3。

表 5-3 路由器端口 IP 地址表

Router-A		Router-B		Router-C	
端口	IP 地址	端口	IP 地址	端口	IP 地址
F0/0	192.168.2.1	F0/0	192.168.3.1	F0/0	192.168.2.2
F0/3	192.168.1.1	F0/3	192.168.1.2	F0/3	192.168.4.1
S0/1 DTE	211.159.127.32				

路由器的基础应用 项目 5

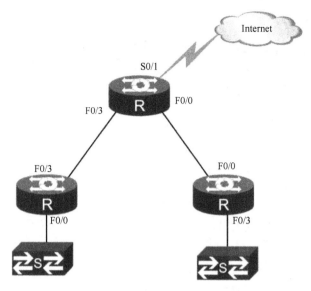

图 5-3 网络拓扑图

任务准备

RIP 的有关介绍详见项目 3 实训四，此处不再赘述。

任务实施

Router-A

1. 设置 Router-A 上所有端口 IP 地址。

```
Router>enable
Router#config
Router_config#hostname Router-A
Router-A_config#interface f0/0
Router-A_config_f0/0#ip address 192.168.2.1 255.255.255.0
Router-A_config_f0/0#exit
Router-A_config#interface f0/3
Router-A_config_f0/3#ip address 192.168.1.1 255.255.255.0
Router-A_config_f0/3#exit
Router-A_config#interface s0/1
Router-A_config_s0/1#ip address 211.159.127.32 255.255.255.0
Router-A_config_s0/1#exit
```

2. 配置 Router-A 内网启用 RIP 协议。

```
Router-A_config#router rip                    //进入 RIP 视图
Router-A_config_rip#version 2                 //设置为 RIPv2
Router-A_config_rip#network 192.168.2.0 255.255.255.0
              //在 192.168.2.0 网段启用 RIP 路由协议
Router-A_config_rip#network 192.168.1.0 255.255.255.0
              //在 192.168.1.0 网段启用 RIP 路由协议
```

3．查看路由器配置。

```
Router-B_config#show running-config
!version 1.3.3H
service timestamps log date
service timestamps debug date
no service password-encryption
hostname Router-A
gbsc group default
interface FastEthernet0/0
 ip address 192.168.2.1 255.255.255.0
 no ip directed-broadcast
interface FastEthernet0/3
 ip address 192.168.1.1 255.255.255.0
 no ip directed-broadcast
interface Serial0/1
ip address 211.159.127.32 255.255.255.0
 no ip directed-broadcast
interface Serial0/2
 no ip address
 no ip directed-broadcast
 physical-layer speed 2048000
interface Async0/0
 no ip address
 no ip directed-broadcast
router rip
 version 2                              //启用RIPv2版本
 network 192.168.1.0 255.255.255.0      //192.168.1.0网段启用RIP路由协议
 network 192.168.2.0 255.255.255.0      //192.168.2.0网段启用RIP路由协议
ip route default 211.159.127.32         //默认路由
```

Router-B

1．设置Router-B所有端口的IP地址。

```
Router>enable
Router#config
Router_config#hostname Router-B
Router-B_config#interface f0/0
Router-B_config_f0/0#ip address 192.168.3.1 255.255.255.0
Router-B_config_f0/0#exit
Router-B_config#interface f0/3
Router-B_config_f0/3#ip address 192.168.1.2 255.255.255.0
Router-B_config_f0/3#exit
```

2．配置RIP协议。

```
Router-B_config#router rip                    //进入RIP视图
Router-B_config_rip#version 2                 //启用RIPv2版本
```

```
Router-B_config_rip#network 192.168.1.0 255.255.255.0
                                    //192.168.1.0 网段启用 RIP 路由协议
Router-B_config_rip#network 192.168.3.0 255.255.255.0
                                    //192.168.3.0 网段启用 RIP 路由协议
```

3. 查看路由器配置。

```
Router-B_config#show running-config
!version 1.3.3H
service timestamps log date
service timestamps debug date
no service password-encryption
hostname Router-B
gbsc group default
interface FastEthernet0/0
 ip address 192.168.3.1 255.255.255.0
 no ip directed-broadcast
interface FastEthernet0/3
 ip address 192.168.1.2 255.255.255.0
 no ip directed-broadcast
interface Serial0/1
 no ip address
 no ip directed-broadcast
interface Serial0/2
 no ip address
 no ip directed-broadcast
 physical-layer speed 2048000
interface Async0/0
 no ip address
 no ip directed-broadcast
router rip
 version 2                              //启用 RIPv2 版本
 network 192.168.1.0 255.255.255.0      //192.168.1.0 网段启用 RIP 路由协议
 network 192.168.3.0 255.255.255.0      //192.168.3.0 网段启用 RIP 路由协议
```

Router-C

1. 如表 5-3 所示，设置 IP 地址。

```
Router>enable
Router#config
Router_config#hostname Router-C
Router-C_config#interface f0/0
Router-C_config_f0/0#ip address 192.168.2.2 255.255.255.0
Router-C_config_f0/0#exit
Router-C_config#interface f0/3
Router-C_config_f0/3#ip address 192.168.4.1 255.255.255.0
Router-C_config_f0/0#exit
```

2. 配置 RIP 协议。

```
Router-C_config#router rip
Router-C_config_rip#version 2
Router-C_config_rip#network 192.168.2.0 255.255.255.0
Router-C_config_rip#network 192.168.4.0 255.255.255.0
```

3. 查看路由器配置。

```
Router-C_config#show running-config
!version 1.3.3H
service timestamps log date
service timestamps debug date
no service password-encryption
hostname Router-C
gbsc group default
interface FastEthernet0/0
 ip address 192.168.2.2 255.255.255.0
 no ip directed-broadcast
interface FastEthernet0/3
 ip address 192.168.4.1 255.255.255.0
 no ip directed-broadcast
interface Serial0/1
 no ip address
 no ip directed-broadcast
interface Serial0/2
 no ip address
 no ip directed-broadcast
interface Async0/0
 no ip address
 no ip directed-broadcast
router rip
 version 2                                    //启用RIPv2
 network 192.168.2.0 255.255.255.0            //192.168.1.0网段启用RIP路由协议
 network 192.168.4.0 255.255.255.0            //192.168.1.0网段启用RIP路由协议
```

所有配置完成后查看路由器中路由表项。

Router-A

```
Router-A_config#show ip route
Codes: C - connected, S - static, R - RIP, B - BGP, BC - BGP connected
     D - DEIGRP, DEX - external DEIGRP, O - OSPF, OIA - OSPF inter area
     ON1 - OSPF NSSA external type 1, ON2 - OSPF NSSA external type 2
     OE1 - OSPF external type 1, OE2 - OSPF external type 2
     DHCP - DHCP type
VRF ID: 0
S      0.0.0.0/0          [1,0] via 211.159.127.32(on Serial0/1)
```

```
C       192.168.1.0/24          is directly connected, FastEthernet0/3
C       192.168.2.0/24          is directly connected, FastEthernet0/0
R       192.168.3.0/24          [120,2] via 192.168.1.2(on FastEthernet0/3)
                                //RIP 路由协议获得 192.168.3.0 网段的路由表项
R       192.168.4.0/24          [120,2] via 192.168.2.2(on FastEthernet0/0)
                                //RIP 路由协议获得 192.168.4.0 网段的路由表项
C       211.159.127.0/24        is directly connected, Serial0/1
```

Router-B

```
Router-B_config#show ip route
Codes: C - connected, S - static, R - RIP, B - BGP, BC - BGP connected
     D - BEIGRP, DEX - external BEIGRP, O - OSPF, OIA - OSPF inter area
     ON1 - OSPF NSSA external type 1, ON2 - OSPF NSSA external type 2
     OE1 - OSPF external type 1, OE2 - OSPF external type 2
     2 DHCP - DHCP type, L1 - IS-IS level-1, L2 - IS-IS level-2
VRF ID: 0
C       192.168.1.0/24          is directly connected, FastEthernet0/3
R       192.168.2.0/24          [120,1] via 192.168.1.1(on FastEthernet0/3)
                                //RIP 路由协议获得 192.168.2.0 网段的路由表项
C       192.168.3.0/24          is directly connected, FastEthernet0/0
R       192.168.4.0/24          [120,2] via 192.168.1.1(on FastEthernet0/3)
                                //RIP 路由协议获得 192.168.4.0 网段的路由表项
```

Router-C

```
Router-C_config#show ip route
Codes: C - connected, S - static, R - RIP, B - BGP, BC - BGP connected
     D - BEIGRP, DEX - external BEIGRP, O - OSPF, OIA - OSPF inter area
     ON1 - OSPF NSSA external type 1, ON2 - OSPF NSSA external type 2
     OE1 - OSPF external type 1, OE2 - OSPF external type 2
     DHCP - DHCP type, L1 - IS-IS level-1, L2 - IS-IS level-2
VRF ID: 0
R       192.168.1.0/24          [120,1] via 192.168.2.1(on FastEthernet0/0)
                                //RIP 路由协议获得 192.168.1.0 网段的路由表项
C       192.168.2.0/24          is directly connected, FastEthernet0/0
R       192.168.3.0/24          [120,2] via 192.168.2.1(on FastEthernet0/0)
                                //RIP 路由协议获得 192.168.3.0 网段的路由表项
C       192.168.4.0/24          is directly connected, FastEthernet0/3
```

实训完成后，利用 "show ip route" 命令，可以看到 3 台路由器的路由表项，通过 RIP 协议获得了内网所有的路由表项（192.168.1.0/24、192.168.2.0/24、192.168.3.0/24 和 192.168.4.0/24）。其中类型 C 为直连路由，类型 R 为从 RIP 路由协议获取的路由表项。

5.4 实训四 路由器 OSPF 配置

任务描述

某公司有 3 台路由器，按如图 5-4 所示拓扑结构进行连接，网络中有多个 IP 地址段，所有路由器配置为动态路由协议 RIP，但以后网络中可能扩大规模并添加大量路由器。由于 RIP 可能在大中型网络中性能较差，网络管理员在路由器上启用 OSPF 路由协议来保证各个网段之间所有 PC 能互联互通。路由器的端口 IP 地址见表 5-4。

表 5-4 路由器端口 IP 地址表

Router-A		Router-B		Router-C	
端口	IP 地址	端口	IP 地址	端口	IP 地址
F0/0	192.168.2.1	F0/0	192.168.3.1	F0/0	192.168.2.2
F0/3	192.168.1.1	F0/3	192.168.1.2	F0/3	192.168.4.1
S0/1 DTE	211.159.127.32				

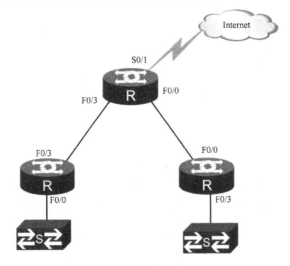

图 5-4 网络拓扑图

任务准备

相关知识详见项目 3 实训五，此处不再赘述。

任务实施

Router-A

1. 如表 5-4 所示，设置 IP 地址。

```
Router>enable
Router#config
Router_config#hostname Router-A
```

```
Router-A_config#interface f0/0
Router-A_config_f0/0#ip address 192.168.2.1 255.255.255.0
Router-A_config_f0/0#exit
Router-A_config#interface f0/3
Router-A_config_f0/3#ip address 192.168.1.1 255.255.255.0
Ro2uter-A_config_f0/3#exit
Router-A_config#interface s0/1
Rou2ter-A_config_s0/1#ip address 211.159.127.32 255.255.255.0
Router-A_config_s0/1#exit
```

2．配置 OSPF 协议。

```
Router-A_config#router ospf 1                  //启动OSPF路由协议
Router-A_config_ospf_1#network 192.168.1.0 255.255.255.0 area 0
                                               //192.168.1.0网段运行OSPF路由协议
Router-A_config_ospf_1#network 192.168.2.0 255.255.255.0 area 0
                                               //192.168.2.0网段运行OSPF路由协议
```

3．查看路由器配置。

```
Router-A_config#show running-config
!version 1.3.3F
service timestamps log date
service timestamps debug date
no service pass2word-encryption
hostname Router-A
gbsc group default
interface FastEthernet0/0
 ip address 192.168.2.1 255.255.255.0
 no ip directed-broadcast
interface FastEthernet0/3
 ip address 192.168.1.1 255.255.255.0
 no ip directed-broadcast
interface Serial0/1
 ip address 211.159.127.32 255.255.255.0
 no ip directed-broadcast
interface Serial0/2
 no ip address
 no ip directed-broadcast
interface Async0/0
 no ip address
 no ip directed-broadcast
router ospf 1                                  //启动OSPF路由协议
 network 192.168.1.0 255.255.255.0 area 0
                                               //192.168.1.0网段运行OSPF路由协议
 network 192.168.2.0 255.255.255.0 area 0
                                               //192.168.2.0网段运行OSPF路由协议
ip route default 211.159.127.32
```

Router-B

1. 如表 5-4 所示，设置 IP 地址。

```
Router>enable
Router#config
Router_config#hostname Router-B
Router-B_config#interface f0/0
Router-B_config_f0/0#ip address 192.168.3.1 255.255.255.0
Router-B_config_f0/0#exit
Router-B_config#interface f0/3
Router-B_config_f0/3#ip address 192.168.1.2 255.255.255.0
Router-B_config_f0/3#exit
```

2. 配置 OSPF 协议。

```
Router-B_config#router ospf 1                    //启动 OSPF 路由协议
Router-B_config_ospf_1#network 192.168.1.0 255.255.255.0 area 0
                                                 //192.168.1.0 网段运行 OSPF 路由协议
Router-B_config_ospf_1#network 192.168.3.0 255.255.255.0 area 0
                                                 //192.168.3.0 网段运行 OSPF 路由协议
```

3. 查看路由器配置。

```
Router-B_config#show running-config
!version 1.3.3H
service timestamps log date
service timestamps debug date
no service password-encryption
hostname Router-B
gbsc group default
interface FastEthernet0/0
 ip address 192.168.3.1 255.255.255.0
 no ip directed-broadcast
interface FastEthernet0/3
 ip address 192.168.1.2 255.255.255.0
 no ip directed-broadcast
interface Serial0/1
 no ip address
 no ip directed-broadcast
interface Serial0/2
 no ip address
 no ip directed-broadcast
 physical-layer speed 2048000
interface Async0/0
 no ip address
 no ip directed-broadcast
router ospf 1                                    //启动 OSPF 路由协议
 network 192.168.1.0 255.255.255.0 area 0
                                                 //192.168.1.0 网段运行 OSPF 路由协议
```

```
network 192.168.3.0 255.255.255.0 area 0
                                         //192.168.3.0网段运行OSPF路由协议
```

Router-C

1. 如表5-4所示，设置IP地址。

```
Router>enable
Router#config
Router_config#hostname Router-C
Router-C_config#interface f0/0
Router-C_config_f0/0#ip address 192.168.2.2 255.255.255.0
Router-C_config_f0/0#exit
Router-C_config#interface f0/3
Router-C_config_f0/3#ip address 192.168.4.1 255.255.255.0
```

2. 配置OSPF协议。

```
Router-C_config#router ospf 1                    //启动OSPF路由协议
Router-C_config_ospf_1#network 192.168.2.0 255.255.255.0 area 0
                                         //192.168.1.0网段运行OSPF路由协议
Router-C_config_ospf_1#network 192.168.4.0 255.255.255.0 area 0
                                         //192.168.4.0网段运行OSPF路由协议
```

3. 查看路由器配置。

```
Router-C_config#show running-config
!version 1.3.3H
service timestamps log date
service timestamps debug date
no service password-encryption
hostname Router-C
gbsc group default
interface FastEthernet0/0
 ip address 192.168.2.2 255.255.255.0
 no ip directed-broadcast
interface FastEthernet0/3
 ip address 192.168.4.1 255.255.255.0
 no ip directed-broadcast
interface Serial0/1
 no ip address
 no ip directed-broadcast
interface Serial0/2
 no ip address
 no ip directed-broadcast
interface Async0/0
 no ip address
 no ip directed-broadcast
router ospf 1                              //启动OSPF路由协议
 network 192.168.2.0 255.255.255.0 area 0
```

```
                                           //192.168.1.0网段运行OSPF路由协议
    network 192.168.4.0 255.255.255.0 area 0
                                           //192.168.4.0网段运行OSPF路由协议
```

所有配置完成后查看路由器中路由表项:

```
Router-A_config#show ip route
Codes: C - connected, S - static, R - RIP, B - BGP, BC - BGP connected
       D - DEIGRP, DEX - external DEIGRP, O - OSPF, OIA - OSPF inter area
       ON1 - OSPF NSSA external type 1, ON2 - OSPF NSSA external type 2
       OE1 - OSPF external type 1, OE2 - OSPF external type 2
       DHCP - DHCP type
VRF ID: 0
S      0.0.0.0/0           [1,0] via 211.159.127.32(on Serial0/1)
C      192.168.1.0/24      is directly connected, FastEthernet0/3
C      192.168.2.0/24      is directly connected, FastEthernet0/0
O      192.168.3.0/24      [110,2] via 192.168.1.2(on FastEthernet0/3)
                           //OSPF路由协议获得192.168.3.0网段的路由表项
O      192.168.4.0/24      [110,2] via 192.168.2.2(on FastEthernet0/0)
                           //OSPF路由协议获得192.168.4.0网段的路由表项
C      211.159.127.0/24    is directly connected, Serial0/1

Router-B_config#show ip route
Codes: C - connected, S - static, R - RIP, B - BGP, BC - BGP connected
       D - BEIGRP, DEX - external BEIGRP, O - OSPF, OIA - OSPF inter area
       ON1 - OSPF NSSA external type 1, ON2 - OSPF NSSA external type 2
       OE1 - OSPF external type 1, OE2 - OSPF external type 2
       DHCP - DHCP type, L1 - IS-IS level-1, L2 - IS-IS level-2
VRF ID: 0
C      192.168.1.0/24      is directly connected, FastEthernet0/3
O      192.168.2.0/24      [110,2] via 192.168.1.1(on FastEthernet0/3)
                           //OSPF路由协议获得192.168.2.0网段的路由表项
C      192.168.3.0/24      is directly connected, FastEthernet0/0
O      192.168.4.0/24      [110,3] via 192.168.1.1(on FastEthernet0/3)
                           //OSPF路由协议获得192.168.4.0网段的路由表项

Router-C_config#show ip route
Codes: C - connected, S - static, R - RIP, B - BGP, BC - BGP connected
       D - BEIGRP, DEX - external BEIGRP, O - OSPF, OIA - OSPF inter area
       ON1 - OSPF NSSA external type 1, ON2 - OSPF NSSA external type 2
       OE1 - OSPF external type 1, OE2 - OSPF external type 2
       DHCP - DHCP type, L1 - IS-IS level-1, L2 - IS-IS level-2
VRF ID: 0
O      192.168.1.0/24      [110,2] via 192.168.2.1(on FastEthernet0/0)
                           //OSPF路由协议获得192.168.1.0网段的路由表项
C      192.168.2.0/24      is directly connected, FastEthernet0/0
O      192.168.3.0/24      [110,3] via 192.168.2.1(on FastEthernet0/0)
```

```
C       192.168.4.0/24         is directly connected, FastEthernet0/3
                               //OSPF 路由协议获得 192.168.4.0 网段的路由表项
```

实训完成后，利用"show ip route"命令，可以看到 3 台路由器的路由表项，通过 OSPF 获得了内网所有的路由表项（192.168.1.0/24、192.168.2.0/24、192.168.3.0/24 和 192.168.4.0/24）。其中类型 C 为直连路由，类型 O 为从 OSPF 路由协议获取的路由表项。

5.5　实训五　路由器串口 PPP 封装

任务描述

某公司有一台路由器通过串口连接到 ISP 路由器，链路带宽为 2Mbps，公司端路由器为 DTE，ISP 端路由器为 DCE。两台路由器端口 IP 地址见表 5-5。

表 5-5　路由器端口 IP 地址

Router-A		Router-B	
端口	IP 地址	端口	IP 地址
S0/2　DCE	192.168.1.1	S0/1　DTE	192.168.1.2

网络拓扑图 5-5 所示。

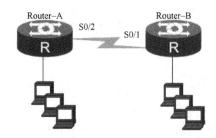

图 5-5　网络拓扑图

任务准备

PPP（Point to Point Protocol）即点对点协议，为在点对点链路上传输多协议数据包提供了一个标准方法。PPP 协议是数据链路层协议，支持点到点的连接，物理层可以是同步电路或异步电路，具有各种 NCP 协议，如 IPCP、IPXCP，更好地支持了网络层协议；具有验证协议 PAP/CHAP，更好地保证了网络的安全性。

任务实施

Router-A

1. 修改路由器名。

```
Router>enable                           //进入特权模式
Router#config                           //进入全局配置模式
Router_config#hostname Router-A         //修改机器名为 "Router-A"
```

2. 进入端口模式，并设置 IP。

```
Router-A_config#interface s0/2                    //进入端口 S0/2
Router-A_config_s0/2#ip address 192.168.1.1 255.255.255.0
                                                  //设置 IP 地址为 "192.168.1.1/24"
```

3. 启用 PPP 封装协议，并设置端口速率为 2048000。

```
Router-A_config_s0/2#encapsulation ppp            //指定封装模式为 "PPP"
Router-A_config_s0/2#physical-layer speed 2048000
                                                  //设置端口速率为 "2048000"
```

4. 查看端口状态。

```
Router-A_config#show interface s0/2               //查看端口 S0/2 状态
Serial0/2 is up, line protocol is down            //物理层 UP，数据链路层 UP
Mode=Sync DCE Speed=2048000
DTR=UP,DSR=UP,RTS=UP,CTS=UP,DCD=UP
MTU 1500 bytes, BW 64 kbit, DLY 2000 usec
Interface address is 192.168.1.1/24
Encapsulation PPP, loopback not set
  Keepalive set(10 sec)
  LCP  Starting configuration exchange
  IPCP Listening -- waiting for remote host to attempt open
  local IP address: 192.168.1.1 remote IP address: 0.0.0.0
  60 second input rate 20 bits/sec, 0 packets/sec!
  60 second output rate 54 bits/sec, 0 packets/sec!
  74 packets input, 1776 bytes, 6 unused_rx, 0 no buffer
  0 input errors, 0 CRC, 0 frame, 0 overrun, 0 ignored, 0 abort
  101 packets output, 2276 bytes, 8 unused_tx, 0 underruns
  error:
  0 clock, 0 grace
  PowerQUICC SCC specific errors:
  0 recv allocb mblk fail     0 recv no buffer
  0 transmitter queue full    0 transmitter hwqueue_full
```

5. 查看路由器配置。

```
Router-A_config#show running-config
!version 1.3.3H
service timestamps log date
service timestamps debug date
no service password-encryption
hostname Router-A
gbsc group default
interface FastEthernet0/0
 no ip address
 no ip directed-broadcast
interface FastEthernet0/3
 no ip address
 no ip directed-broadcast
```

```
interface Serial0/1
 no ip address
 no ip directed-broadcast
interface Serial0/2
 ip address 192.168.1.1 255.255.255.0
 no ip directed-broadcast
 encapsulation ppp                          //封装协议为 PPP
 physical-layer speed 2048000               //端口速率为"2048000"
interface Async0/0
 no ip address
 no ip directed-broadcast
```

Router-B

1. 修改路由器名。

```
Router>enable                               //进入特权模式
Router#config                               //进入全局配置模式
Router_config#hostname Router-B             //修改机器名为"Router-B"
```

2. 进入端口模式,并设置 IP。

```
Router-B_config#interface s0/1              //进入端口 S0/1
Router-B_config_s0/2#ip address 192.168.1.2 255.255.255.0
                                            //设置 IP 地址为"192.168.1.2/24"
```

3. 启用 PPP 封装协议。

```
Router-B_config_s0/2#encapsulation ppp      //指定封装协议为"PPP"
```

4. 查看端口状态。

```
Router-B_config#show interface s0/1         //查看端口 S0/1 状态
Serial0/1 is up, line protocol is up
Mode=Sync DTE
DTR=UP,DSR=UP,RTS=UP,CTS=UP,DCD=UP
MTU 1500 bytes, BW 64 kbit, DLY 2000 usec
Interface address is 192.168.1.2/24
Encapsulation PPP, loopback not set
Keepalive set(10 sec)
LCP Opened
IPCP Opened
local IP address: 192.168.1.2  remote IP address: 192.168.1.1
60 second input rate 57 bits/sec, 0 packets/sec!
60 second output rate 20 bits/sec, 0 packets/sec!
207 packets input, 4384 bytes, 5 unused_rx, 0 no buffer
0 input errors, 0 CRC, 0 frame, 0 overrun, 0 ignored, 0 abort
110 packets output, 2612 bytes, 8 unused_tx, 0 underruns
error:
0 clock, 0 grace
```

```
PowerQUICC SCC specific errors:
0 recv allocb mblk fail    0 recv no buffer
0 transmitter queue full   0 transmitter hwqueue_full
```

5. 查看路由器配置

```
Router-B_config#show running-config
!version 1.3.3H
service timestamps log date
service timestamps debug date
no service password-encryption
hostname Router-B
gbsc group default
interface FastEthernet0/0
no ip address
no ip directed-broadcast
interface FastEthernet0/3
no ip address
no ip directed-broadcast
interface Serial0/1
ip address 192.168.1.2 255.255.255.0
no ip directed-broadcast
encapsulation ppp                          //封装协议为PPP
interface Serial0/2
no ip address
no ip directed-broadcast
interface Async0/0
no ip address
no ip directed-broadcast
```

在实验过程中，当两台路由器正常加电，并且线缆连接正常时，端口物理层就会 UP 状态。神州数码路由器的串口默认情况下没有启用任何协议，当 Router-A 配置后，而 Router-B 没有配置时，Router-A 的 Serial0/1 接口是 DOWN 状态。当两台路由器都配置正确后，路由器间链路的数据链路层才会 UP，用户可以通过 ping 命令测试网络联通性。

5.6 实训六 PPP 封装协议 CHAP 认证

任务描述

某公司有一台路由器通过串口连接到 ISP 路由器，链路带宽为 2Mbps，公司端路由器为 DTE，ISP 端路由器为 DCE，为了保护两台路由器间链路安全性，在两台路由器间的链路上启用 CHAP 认证。两台路由器端口 IP 地址见表 5-6。

路由器的基础应用 项目 5

表 5-6 路由器端口 IP 地址

Router-A		Router-B	
端口	IP 地址	端口	IP 地址
S0/2 DCE	192.168.1.1	S0/1 DTE	192.168.1.2
账号	密码	账号	密码
RouterA	123123	RouterB	123123

网络拓扑图如图 5-6 所示。

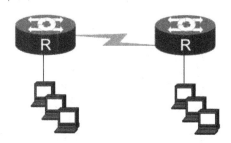

图 5-6 网络拓扑图

任务准备

CHAP（Challenge Handshake Authentication Protocol）是 PPP 询问握手认证协议。此协议可通过三次握手周期性地校验对端的身份，可在初始链路建立时完成，在链路建立之后重复进行。通过递增改变的标识符和可变的询问值，可防止来自端点的重放攻击，限制暴露于单个攻击的时间。

任务实施

Router-A

1. 修改路由器名。

```
Router>enable                              //进入特权模式
Router#config                              //进入全局配置模式
Router_config#hostname Router-A            //修改机器名为"Router-A"
```

2. 进入端口模式，并设置 IP。

```
Router-A_config#interface s0/2             //进入端口 S0/2
Router-A_config_s0/2#ip address 192.168.1.1 255.255.255.0
                                           //设置 IP 地址为"192.168.1.1/24"
```

3. 启用 PPP 封装协议，并设置端口速率为 2048000。

```
Router-A_config_s0/2#encapsulation ppp     //指定封装协议为"PPP"
Router-A_config_s0/2#physical-layer speed 2048000
                                           //设置端口速率为"2048000"
```

4、设置 PPP 封装协议为 CHAP 认证，并设置主机名为"RouterA"。

```
Router-A_config_s0/2#ppp authentication chap   //设置 PPP 封装协议为"CHAP"认证
Router-A_config_s0/2#ppp chap hostname RouterA
```

//设置路由器用户名为"RouterA"
```
Router-A_config_s0/2# ppp chap password 123123    //设置路由器密码为123123
```

5. 查看端口状态。

```
Router-A_config#show interface s0/2
Serial0/2 is up, line protocol is down          //物理端口UP, 链路协议DOWN
 Mode=Sync DCE Speed=2048000
  DTR=UP,DSR=UP,RTS=UP,CTS=UP,DCD=UP
  MTU 1500 bytes, BW 64 kbit, DLY 2000 usec
  Interface address is 192.168.1.1/24
  Encapsulation PPP, loopback not set
  Keepalive set(10 sec)
  LCP Opened
  CHAP Listening -- waiting for remote host to attempt open,  Message:
'CHAP: Inv
  alid Response'
  IPCP Listening -- waiting for remote host to attempt open
       local IP address: 192.168.1.1  remote IP address: 0.0.0.0
  60 second input rate 218 bits/sec, 0 packets/sec!
  60 second output rate 225 bits/sec, 0 packets/sec!
    634 packets input, 16494 bytes, 4 unused_rx, 0 no buffer
    0 input errors, 0 CRC, 0 frame, 0 overrun, 0 ignored, 0 abort
    778 packets output, 19710 bytes, 8 unused_tx, 0 underruns
  error:
    0 clock, 0 grace
  PowerQUICC SCC specific errors:
    0 recv allocb mblk fail    0 recv no buffer
    0 transmitter queue full    0 transmitter hwqueue_full
```

6. 查看路由器配置。

```
Router-A_config#show running-config
!version 1.3.3H
service timestamps log date
service timestamps debug date
no service password-encryption
hostname Router-A
gbsc group default
aaa authentication ppp default local
username RouterB password 0 123123
interface FastEthernet0/0
 no ip address
 no ip directed-broadcast
interface FastEthernet0/3
 no ip address
 no ip directed-broadcast
interface Serial0/1
 no ip address
 no ip directed-broadcast
```

```
interface Serial0/2
 ip address 192.168.1.1 255.255.255.0
 no ip directed-broadcast
 encapsulation ppp
 ppp authentication chap                  //chap 认证方式
 ppp chap hostname RouterA                //chap 认证用户名
 ppp chap password 123123                 //chap 认证密码
 physical-layer speed 2048000
interface Async0/0
 no ip address
 no ip directed-broadcast
```

Router-B

1. 修改路由器名。

```
Router>enable              //进入特权模式
Router#config              //进入全局配置模式
Router_config#hostname Router-B          //修改机器名为"Router-B"
```

2. 进入端口模式，并设置 IP。

```
Router-B_config#interface s0/1           //进入端口 S0/1
Router-B_config_s0/2#ip address 192.168.1.2 255.255.255.0
                                         //设置 IP 地址为"192.168.1.2/24"
```

3. 启用 PPP 封装协议。

```
Router-B_config_s0/2#encapsulation ppp   //启用 PPP 封装协议
```

4. 设置 PPP 封装协议为 CHAP 认证，并设置主机名为"RouterB"。

```
Router-B_config_s0/2#ppp authentication chap
                                         //设置 PPP 封装协议为"CHAP"认证
Router-B_config_s0/2#ppp chap hostname RouterB
                                         //设置路由器用户名为"RouterB"
Router-B_config_s0/2# ppp chap password 123123
                                         //设置路由器密码为123123
```

5. 查看端口状态。

```
Router-B_config#show interface s0/1      //查看端口 S0/1 状态
Serial0/1 is up, line protocol is up     //物理端口 UP，链路协议 UP
 Mode=Sync DTE
  DTR=UP,DSR=UP,RTS=UP,CTS=UP,DCD=UP
  MTU 1500 bytes, BW 64 kbit, DLY 2000 usec
  Interface address is 192.168.1.2/24
  Encapsulation PPP, loopback not set
  Keepalive set(10 sec)
  LCP Opened
  CHAP Opened,  Message:'Welcome to Digital China Networks Limited Router'
  IPCP Opened
```

```
     local IP address: 192.168.1.2  remote IP address: 192.168.1.1
  60 second input rate 89 bits/sec, 0 packets/sec!
  60 second output rate 122 bits/sec, 0 packets/sec!
    106 packets input, 2917 bytes, 4 unused_rx, 0 no buffer
    0 input errors, 0 CRC, 0 frame, 0 overrun, 0 ignored, 0 abort
    151 packets output, 4113 bytes, 8 unused_tx, 0 underruns
  error:
    0 clock, 0 grace
  PowerQUICC SCC specific errors:
    0 recv allocb mblk fail      0 recv no buffer
    0 transmitter queue full     0 transmitter hwqueue_full
```

6. 查看路由器配置。

```
Router-B_config#show running-config
!version 1.3.3H
service timestamps log date
service timestamps debug date
no service password-encryption
hostname Router-B
gbsc group default
aaa authentication ppp default local
username RouterA password 0 123123
interface FastEthernet0/0
 no ip address
 no ip directed-broadcast
interface FastEthernet0/3
 no ip address
 no ip directed-broadcast
interface Serial0/1
 ip address 192.168.1.2 255.255.255.0
 no ip directed-broadcast
 encapsulation ppp
 ppp authentication chap              //chap 认证方式
 ppp chap hostname RouterB            //chap 认证用户名
 ppp chap password 123123             //chap 认证密码
interface Serial0/2
 no ip address
 no ip directed-broadcast
interface Async0/0
 no ip address
 no ip directed-broadcast
```

在实验过程中，当两台路由器正常加电，并且线缆连接正常时，接口物理层就会 UP 状态。Router-A 配置完成后，Router-B 端尚未进行配置，所以两端 PPP 协商失败，造成 Router-A 的接口 Serial0/1 数据链路层状态为 DOWN。当两台路由器配置完成后，PPP 协商通过后，Serial0/1 接口数据链路层状态变为 UP。

5.7 实训七 NAT 网络地址转换

任务描述

某公司外网 IP 地址为"211.159.127.32",内网 IP 地址为"192.168.1.1",内网仅有一个网段为"192.168.1.0/24",因为内网为私有 IP 地址,而 Internet 主机不认识私有 IP 地址,所以需要在路由器上启用 NAT,来使路由器把内部本地地址映射成内部全局地址(此处映射为 211.159.127.32),进而能正常访问 Internet。

网络拓扑图如图 5-7 所示。

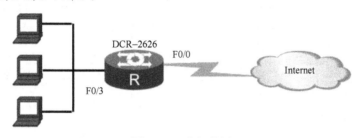

图 5-7 网络拓扑图

任务准备

公有 IP 地址:公有 IP 地址是指在互联网上全球唯一的 IP 地址。

私有 IP 地址:私有 IP 地址是指内部网络或主机的 IP 地址,私有网络预留出了三个 IP 地址段,如下所示。

- A 类:10.0.0.0 ~ 10.255.255.255
- B 类:172.16.0.0 ~ 172.31.255.255
- C 类:192.168.0.0 ~ 192.168.255.255

NAT(Network Address Translation)即网络地址转换,属接入广域网技术,是一种将私有(保留)IP 地址转化为合法 IP 地址的转换技术,它被广泛应用于各种类型 Internet 接入方式和各种类型的网络中。NAT 不仅解决了公网 IP 地址不足的问题,隐藏并保护网络内部的计算机,而且还能够有效地避免来自网络外部的攻击。

NAT 的实现方式有三种,即静态转换、动态转换和端口多路复用。

静态转换(Static Nat)是指将内部网络的私有 IP 地址转换为公有 IP 地址,IP 地址对是一对一的,某个私有 IP 地址只转换为某个公有 IP 地址。

动态转换(Dynamic Nat)是指将内部网络的私有 IP 地址转换为公用 IP 地址时,IP 地址是路由器从合法外部地址集中随机分配的地址,所有被授权访问 Internet 的私有 IP 地址可随机转换为任何指定的合法 IP 地址。

端口多路复用(Network address Port Translation,NAPT) 即端口地址转换,是指改变外出数据包的源端口和 IP 地址并进行端口转换。内部网络的所有主机均可共享一个合法外部 IP 地址实现对 Internet 的访问,从而可以最大限度地节约 IP 地址资源。目前网络中应用最多的就是端口多路复用方式。

内部本地地址：内网中设备所使用的地址，一般为私有地址。

内部全局地址：在路由器或防火墙上设置的公有地址，一般由 ISP 提供，把内网 IP 转换为此地址，进而内网能和外网通信。

任务实施

1. 将路由器端口设置相应 IP 地址，并把端口设置为 NAT 内部端口和 NAT 外部端口。

```
Router>enable
Router#config
Router_config#interface f0/0
Router_config_f0/0#ip address 211.159.127.32 255.255.255.0
Router_config_f0/0#ip nat outside              //设置为 NAT 外部端口
Router_config_f0/0#exit
Router_config#interface f0/3
Router_config_f0/3# ip address 192.168.1.1 255.255.255.0
Router_config_f0/3#ip nat inside               //设置为 NAT 内部端口
Router_config_f0/3#exit
```

2. 创建访问控制列表。

```
Router_config#ip access-list standard 123       //建立标准访问列表，列表名为 123
Router_config_std_nacl#permit any               //允许所有 IP 地址
Router_config_std_nacl#exit
```

3. 创建 NAT。

```
Router_config#ip nat insides source list 123 interface f0/0
                   //启用 NAPT，允许内网 IP 地址转换为 F0/0 端口 IP 地址
```

4. 查看路由器配置。

```
Router_config#show running-config
!version 1.3.3H
service timestamps log date
service timestamps debug date
no service password-encryption
gbsc group default
interface FastEthernet0/0
 ip address 211.159.127.32 255.255.255.0
 no ip directed-broadcast
 ip nat outside                                 //NAT 外部端口
interface FastEthernet0/3
 ip address 192.168.1.1 255.255.255.0
 no ip directed-broadcast
 ip nat inside                                  //NAT 内部端口
interface Serial0/1
 no ip address
 no ip directed-broadcast
interface Serial0/2
 no ip address
```

```
 no ip directed-broadcast
interface Async0/0
 no ip address
 no ip directed-broadcast
 ip access-list standard 123            //标准访问列表名为123
  permit any                            //允许所有IP地址
 ip nat inside source list 123 interface FastEthernet0/0
                    //启用NAPT，允许内网IP地址转换为F0/0端口IP地址
```

在路由器上设置好NAT后，内网用户只要设置了正确的网关和DNS，就能访问Internet。上网过程中，路由器把内部本地地址转换为内部全局地址，可以通过命令"show ip nat translation"显示内部本地地址和内部全局地址的对应。命令显示篇幅过大故在此不再赘述。

项目 6

认识防火墙

教学目标

通过本章的学习,学生可以了解防火墙的基础配置命令以及针对基础配置的方式和方法。为接下来的防火墙的具体应用配置打好基础。

能力目标

了解防火墙在网络中的作用
熟悉防火墙的基础配置方式

知识目标

熟悉防火墙的端口分类
熟悉防火墙的管理模式

主要教学内容

防火墙的基础知识
防火墙配置文件的管理

认识防火墙 项目 6

6.1　　实训一　防火墙外观与端口

本实训中所采用的防火墙为神州数码 DCFW-1800S-H-V2，采用神州数码自主开发的 64 位实时安全操作系统 DCFOS，具备强大的并行处理能力。DCFOS 采用专利的多处理器全并行架构，和常见的处理器或 NP/ASIC 只负责三层包转发的架构不同，实现了从网络层到应用层的全并行处理。因此，神州数码 DCFW-1800 系列安全网关产品较业界其他的或 NP/ASIC 系统在同档的硬件配置下能使性能提升 5 倍以上，为同时开启多项防护功能奠定了性能基础，突破了传统安全网关在功能和性能上无法两全的局限。

随着网络的快速发展，越来越多的应用都建立在 HTTP/HTTPS 等应用层协议之上。新的安全威胁也随之嵌入到应用之中，而传统基于状态检测的安全网关只能依据端口或协议去设置安全策略，根本无法识别应用，更谈不上安全防护。神州数码新一代安全网关可根据应用的行为和特征实现对该应用的识别和控制，而不依赖于端口或协议，即使加密过的数据流也能应付自如。

神州数码系列安全网关提供全方位的安全防护，包括病毒过滤、入侵防御、内容过滤、网页访问控制和应用流量整形等功能，可防范病毒、间谍软件、蠕虫、木马等网络攻击。关键字过滤和基于超过 2000 万域名的 Web 页面分类数据库可以帮助管理员轻松设置工作时间禁止访问的网页，提高工作效率和控制对不良网站的访问。病毒特征库、攻击库、URL 库可以通过网络服务实时下载，确保对新爆发的病毒、攻击、新网页的及时响应。

DCFOS 的应用和身份识别功能能够满足越来越多的深度安全需求。基于身份和角色的管理（RBNS）让网络配置更加直观和精细化。不同的用户甚至同一用户在不同的地点或时间都可以有不同的管理策略。用户访问的内容也可以记录在本机存储模块或专用服务器中，通过用户的名称审阅相关记录使查找更简单。基于角色的管理模式主要包含基于"人"的访问控制、基于"人"的网络资源（服务）的分配、基于"人"的日志审计三方面。基于角色的管理模式可以通过对访问者身份审核和确认，确定访问者的访问权限，分配相应的网络资源。在技术上可避免 IP 盗用或者 PC 终端被盗用引发的数据泄露等问题。

神州数码安全网关部分型号产品采用模块化设计，支持三种类型的扩展模块：接口扩展模块、应用处理扩展模块和存储扩展模块。不同扩展模块的应用可以实现用户不同应用环境的需求，进而充分地保护用户投资；通过增加接口扩展模块提高设备的连接性，使设备不会因为网络带宽或应用系统的升级而过时；通过增加应用处理扩展模块提高本机应用处理能力，让应用处理不再成为性能瓶颈；通过增加存储扩展模块提供实时记录日志信息（NAT 日志、Web 访问日志、Web 内容审计日志等）功能，使外置高性能日志服务器的使用成为现实，以适应校园网和小区宽带等大流量的应用场景。

神州数码 DCFW-1800 系列安全网关采用一体化机箱设计，可以安装在 19 英寸标准机柜中使用，也可以独立卧式使用。

本实训中采用神州数码 DCFW-1800S-H-V2，其前面板布局从左到右依次为各种指示灯、1 个配置端口、1 个 USB 端口和 5 个以太网端口，后面板有电源端口和接地柱，其前面板示意图如图 6-1 所示。

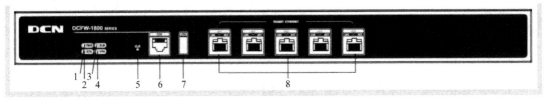

图 6-1　防火墙前面板示意图

防火墙的前面板示意图中序号代表的标示见表 6-1。

表 6-1　防火墙标识及说明

序　号	标识及说明	序　号	标识及说明
1	PWR：电源指示灯	2	STA：状态指示灯
3	ALM：警告指示灯	4	VPN：VPN 状态指示灯
5	CLR：CLR 按键	6	CON：配置口
7	USB：USB 端口	8	E0/0～E0/4：以太网端口

防火墙各个指示灯颜色及状态所代表的含义见表 6-2。

表 6-2　指示灯颜色及状态说明

指　示　灯	颜色/状态	含　义
PWR	绿色常亮	系统电源工作正常
	橙色常亮	电源工作异常
	红色常亮	电源工作异常，此时系统进入关闭状态
STA	熄灭	系统没有供电或处于关闭状态
	绿色常亮	系统处于启动状态
	绿色闪烁	系统已启动并且正常工作
ALM	红色常亮	系统启动失败或者系统异常
	橙色闪烁	系统正在使用试用许可证
	绿色闪烁	系统处于等待状态
	熄灭	系统正常
	橙色	系统的试用许可证已过期，无合法许可证
LNK	绿色常亮	端口与对端设备通过网线或光纤连接且连接正常
	熄灭	端口与对端设备无连接或端口连接失败
ACT	黄色闪烁	端口处于收发数据状态
	熄灭	端口无数据传输

神州数码 DCFW-1800S-H-V2 安全网关必须在室内使用，为保证安全网关正常工作和延长使用寿命，安装场所应该满足下列要求：

● 灰尘粒子直径 ≥5 μm，灰尘粒子小于 3×10^4 粒/m^3。

● 为防止静电损伤，应做到设备良好接地，机箱后面的接地保护螺丝与接地线连通。

● 抗电磁干扰的要求：应做到对供电系统采取有效的防电网干扰措施；安全网关的工作接地最好不要与电力设备的接地装置或防雷接地装置合用，并尽可能相距远一些；远离强功率无线电发射台、雷达发射台和高频大电流设备。

6.2 实训二 防火墙管理模式

神州数码 DCFW-1800S-H-V2 所采用的操作系统为 DCFOS（神州数码防火墙操作系统），此设备操作系统支持本地和远程环两种境配置方法，可以通过命令行和 WebUI（Web 图形接口）两种方式进行配置。

命令行方式：

1. 用标准的 RS-232 电缆将 PC 的串口与安全网关的 CON 口连接起来，如图 6-2 所示。

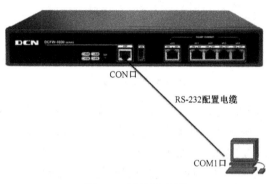

图 6-2　连接示意图

2. 在计算机上运行终端仿真程序（Windows 下的超级终端）建立与安全网关的连接，并按照图 6-3 配置 COM1 口参数。

图 6-3　COM1 口属性配置

3. 给安全网关 DCFW-1800S-H-V2 上电，安全网关会进行自检，进行系统初始化配置。如果系统启动成功，会出现登录提示"login:"，在登录提示后输入默认管理员名称"admin"并敲回车键，界面出现密码提示"password"，输入默认密码"admin"并敲回车键，此时用户便成功登录并且进入 CLI 配置界面，如图 6-4 所示。

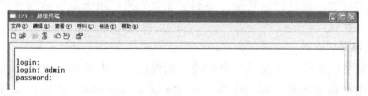

图 6-4　超级终端中输入用户名和密码

4. 用命令对安全网关进行配置或者查看安全网关的运行状态,使用命令时可随时键入"?"寻求帮助,如图 6-5 所示。

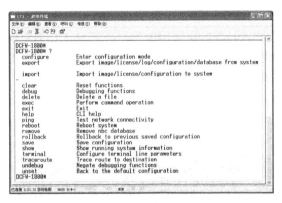

图 6-5　防火墙命令

DCFOS 命令行有不同级别的命令模式,一些命令只有在特定的命令模式下才可使用。例如,只有在相应的配置模式下,才可以输入并执行配置命令,这样也可以防止意外破坏已有的配置。不同的命令模式都有其相应的 CLI 提示符。

执行模式：用户进入到 CLI 时的模式是执行模式。执行模式允许用户使用其权限级别允许的所有的设置选项。该模式的提示符包含了一个数字符号（#）：DCFW-1800#。

全局配置模式：全局配置模式允许用户修改安全网关的配置参数。用户在执行模式下,输入"configure"命令,可进入全局配置模式。该模式的提示符为：DCFW-1800 (config)#。

子模块配置模式：安全网关的不同模块功能需要在其对应的命令行子模块模式下进行配置。用户在全局配置模式输入特定的命令可以进入相应的子模块配置模式。例如,运行"interface ethernet0/0"命令进入 Ethernet0/0 端口配置模式,此时的提示符变更为：DCFW-1800 (config-if-eth0/0)#。

CLI 命令模式切换用户登录到安全网关就进入到 CLI 的执行模式。用户可以通过不同的命令在各种命令模式之间进行切换。表 6-3 列出 CLI 的模式切换命令。

表 6-3　命令行切换命令

模　　式	命　　令
执行模式到全局配置模式	Configure
全局配置模式到子模块配置模式	不同功能使用不同的命令进入各自的命令配置模式
退回到上一级命令模式	Exit
从任何模式退回到执行模式	End

DCFOS 命令行能完成神州数码防火墙所有的功能,但其命令与神州数码交换机和路由器相比较有较大差别,而且比较复杂,不利于配置。所以防火墙日常的管理任务大部分在图形管理界面下完成。

WebUI 配置环境：

神州数码安全网关除可以使用命令行进行配置以外,还提供 WebUI 界面,使用户能够更简便与直观地对设备进行管理与配置。安全网关的 Ethernet0/0 端口配有默认 IP 地址

192.168.1.1/24，并且该端口的各种管理功能均为开启状态。初次使用安全网关时，用户可以通过该端口访问安全网关的 WebUI 页面。按照以下步骤可以搭建 WebUI 配置环境。

1. 将管理 PC 的 IP 地址设置为与 192.168.1.1/24 同网段的 IP 地址，并且用网线将管理 PC 与安全网关的 Ethernet0/0 端口进行连接。

2. 在管理 PC 的 Web 浏览器中访问地址 http://192.168.1.1 并按回车键，出现如图 6-6 所示登录页面。

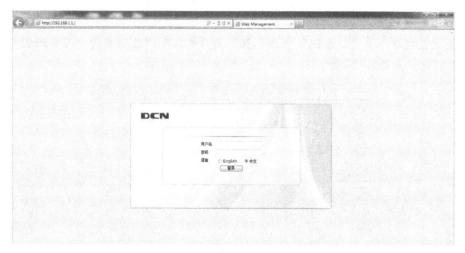

图 6-6　防火墙 WebUI 登录界面

3. 在网页中输入默认管理员名称"admin"和密码"admin"，语言选择"中文"，然后单击"登录"按钮，就可以登录防火墙图形界面，界面如图 6-7 所示。

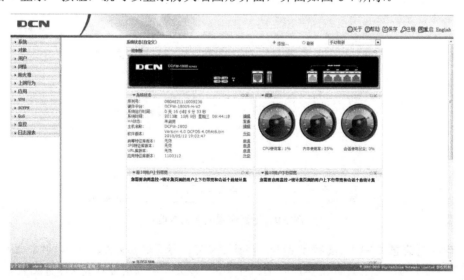

图 6-7　防火墙系统状态界面

这样用户就可以通过防火墙 WebUI 界面，配置防火墙的大部分功能，但一部分功能的实现必须在命令行下，例如，OSPF 协议和负载均衡必须在命令行下配置。

6.3 实训三 管理防火墙配置文件

防火墙的配置信息都被保存在系统的配置文件中,用户通过运行相应的命令或者访问相应的 WebUI 页面查看安全网关的各种配置信息,例如防火墙的初始配置信息和当前配置信息等。配置文件以命令行的格式保存配置信息,并且也以这种格式显示配置信息。

WebUI 配置环境下管理配置文件:

查看配置信息:单击"系统"中"配置"选项卡,可以直接查看"当前配置"中的防火墙配置信息,如图 6-8 所示。

图 6-8 防火墙配置管理界面

单击"下载"按钮,可出现如图 6-9 所示界面,将配置文件保存到本地进行查看或直接用记事本打开配置文件。

图 6-9 下载配置文件到本地

保存配置信息:在防火墙所有 WebUI 界面中都可以保存配置文件,例如,单击上图右上角"保存"按钮,就可以出现如图 6-10 界面,并单击"确定"按钮就可以成功保存配置文件。

图 6-10 配置文件名称

删除配置信息：如果需要删除防火墙配置文件，恢复出厂设置，以删除 current 配置文件为例，选择"配置记录"下的 current 条目后的垃圾桶标识，就会出现如图 6-11 所示界面，单击"确定"按钮。

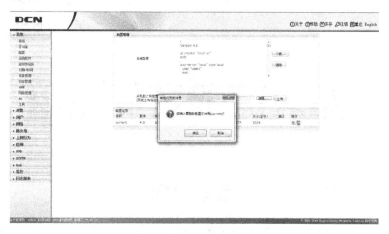

图 6-11　删除配置文件

单击"清除"按钮，那么就会出现如图 6-12 所示界面，单击"确定"按钮，就可以重启系统，并恢复到出厂设置。

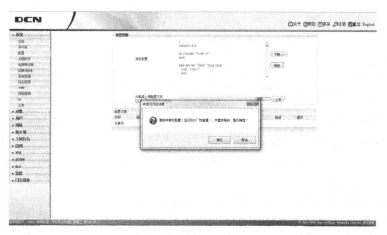

图 6-12　恢复出厂设置

项目 7

防火墙的具体应用

教学目标

通过本章的学习,学生可以了解防火墙的基础配置以及针对这些基础配置的方式和方法,掌握常用的防火墙配置的过程和思维模式,能针对常见防火墙应用进行配置。

能力目标

了解防火墙的常见应用
熟悉防火墙的配置方法
掌握防火墙常见应用的配置

知识目标

熟悉防火墙相关应用的原理
熟悉防火墙的配置界面
熟悉防火墙的相关安全配置

主要教学内容

防火墙安全应用的配置
防火墙常见应用的配置
防火墙辅助功能的配置

7.1 实训一 配置防火墙网络接口 IP 及 zone

任务描述

某网络公司为了保证业务的安全，对公司网络进行了重新规划，公司内网 IP 地址段为 192.168.1.0/24，网络边缘使用神州数码 DCFW-1800 作为公司的防火墙，并在电信申请了 100Mb/s 专线，并分配公网 IP 地址为 211.147.176.231/24，网关为 211.147.176.1。公司网络管理员由于对防火墙的特性还在逐步的摸索过程中，为了保证内网用户能正常上网，对防火墙进行了初步配置。

网络拓扑结构如图 7-1 所示。

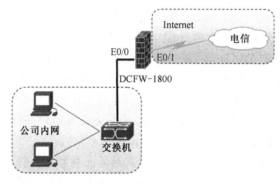

图 7-1 网络拓扑图

防火墙接口 IP 地址分配见表 7-1。

表 7-1 防火墙接口 IP 分配表

防火墙接口号	IP 地址
E0/0	192.168.1.1/24
E0/1	211.147.176.231/24

任务准备

神州数码防火墙与路由器结构的组成部分有些差异，DCFOS（神州数码防火墙操作系统）中的基本组成部分包括：接口、安全域、VSwitch、VRouter、策略以及 VPN。

安全域：安全域将网络划分为不同信任等级部分，例如 trust（通常为内网等可信任部分）、untrust（通常为互联网等存在安全威胁的不可信任部分）等。将配置的策略规则应用到安全域上后，安全网关就能够对出入安全域的流量进行管理和控制。DCFOS 提供 8 个预定义安全域，分别是：trust、untrust、dmz、L2-trust、L2-untrust、L2-dmz、VPNHub 和 HA。

接口：由于防火墙的策略中是基于安全域进行流量的管理和控制，为使流量能够流入和流出某个安全域，必须将接口绑定到某安全域。如果安全域是三层安全域，还需要为接口配置 IP 地址，并且配置相应的策略规则，允许流量在不同安全域中的接口之间传输。多个接口可以被绑定到一个安全域，但是一个接口不能被绑定到多个安全域。

VSwitch：VSwitch（Virtual Switch）即虚拟交换机，具有交换机功能。VSwitch 工作

在二层，将二层安全域绑定到 VSwitch 上后，绑定到安全域的接口也被绑定到该 VSwitch 上。DCFOS 有一个默认的 VSwitch，名为 VSwitch1，默认情况下，二层安全域都会被绑定到 VSwtich1 中。用户可以根据需要创建其他 VSwitch，绑定二层安全域到不同 VSwitch 中。一个 VSwitch 就是一个二层转发域，每个 VSwitch 都有自己独立的 MAC 地址表，因此设备的二层转发在 VSwitch 中实现。并且，流量可以通过 VSwitch 接口，实现二层与三层之间的转发。

VRouter：VRouter（Virtual Router）即虚拟路由器，在 DCFOS 系统中简称为 VR。VRouter 具有路由器功能，不同 VR 拥有各自独立的路由表。系统中有一个默认 VR，名为 trust-vr，默认情况下，所有三层安全域都将会自动绑定到 trust-vr 上。

策略：策略实现防火墙保证网络安全的功能。策略通过策略规则决定从一个安全域到另一个安全域的哪些流量该被允许，哪些流量该被拒绝。默认情况下，所有通过安全网关的流量都是被拒绝的，用户可以根据需要创建策略规则，允许特定的流量在不同安全域之间或者安全域内通过。例如，允许从 trust 域发起到 untrust 域的所有类型流量通过，或者只允许从 untrust 域发起到 DMZ 域的某种特定应用类型的流量通过。

接口绑定到安全域：绑定到二层安全域的接口为二层接口，绑定到三层安全域的接口为三层接口。一个接口只能绑定到一个安全域，主接口与子接口可以分别属于不同的安全域。

安全域绑定到 VSwitch 或者 VRouter：二层安全域绑定到 VSwitch，三层安全域绑定到 VRouter。一个安全域只能绑定到一个 VSwtich 或者 VRouter。

NAT：NAT（Network Address Translation，网络地址转换）是将 IP 数据包头中的 IP 地址转换为另一个 IP 地址的过程。NAT 主要用于实现私有网络访问公共网络的功能。通过使用少量的公有 IP 地址代表较多的私有 IP 地址，有助于减缓可用 IP 地址空间的枯竭。NAT 不仅完美地解决了公网 IP 地址不足的问题，而且还能够有效地避免来自网络外部的攻击，隐藏并保护网络内部的计算机。

任务实施

1. 设置管理 PC 的 IP 地址为"192.168.1.2"，并通过网线连接防火墙 E0/0 接口，打开 IE 浏览器访问防火墙内网管理 IP 地址"192.168.1.1"，登录防火墙图形界面，选择"网络"菜单中"接口"，出现如图 7-2 所示界面。Ethernet 0/0 的默认 IP 地址为 192.168.1.1，所属安全域为"trust"域，就不再进行配置。

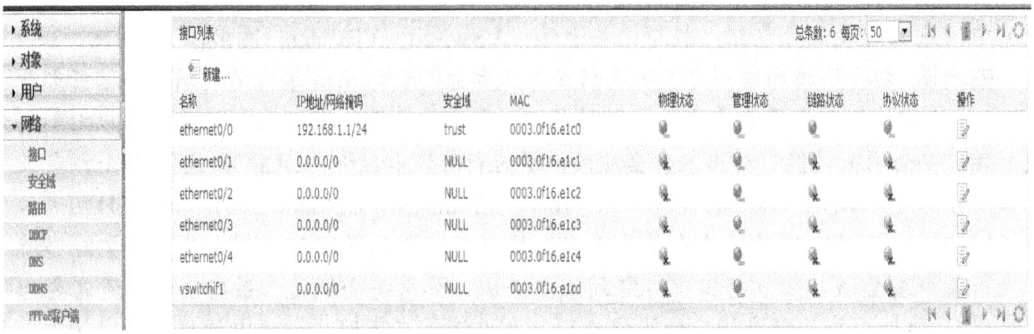

图 7-2 网络接口列表

2. 单击 Ethernet 0/1 接口的"操作"按钮，出现如图 7-3 所示界面。

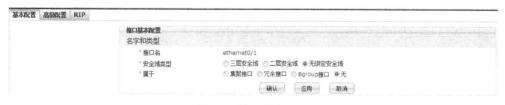

图 7-3　接口基本配置

3．安全域类型选择"三层安全域"，安全域选择"untrust"，设置 IP 为"221.147.176.231/24"，如图 7-4 所示，单击"确认"按钮。

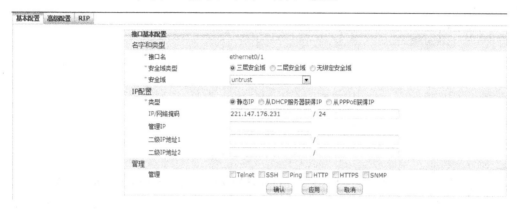

图 7-4　配置 Ethernet0/1 接口

4．选择"防火墙"菜单的"策略"选项，出现如图 7-5 所示界面。

图 7-5　策略列表

5．单击"新建"，选择源安全域"trust"到目的安全域"untrust"，服务簿改为"any"，行为为"允许"，如图 7-6 所示，单击"确认"按钮新建一条内网到外网允许所有数据的防火墙策略。

图 7-6　策略基本配置

6. 选择"防火墙"菜单中"NAT"下的"源 NAT",出现如图 7-7 所示界面。

图 7-7 源 NAT 列表

7. 单击"创建",选择源地址为"any",出接口为外网端口"ethernet0/1",行为为"NAT(出接口 IP)",如图 7-8 所示,单击"确认"按钮,配置内网到外网的路由类型为 NAT。

图 7-8 源 NAT 基本配置

8. 选择"网络"菜单中"路由"下的"目的路由",出现如图 7-9 所示界面。

图 7-9 目的路由

9. 单击"新建",设置目的 IP 为"0.0.0.0",子网掩码为"0.0.0.0",下一跳选择"网关",网关设置为"211.147.176.1",如图 7-10 所示,单击"确认"按钮,配置网络的默认路由。

图 7-10 目的路由配置

10. 在客户机上进行测试，访问站点 http://www.baidu.com 成功，如图 7-11 所示，同样客户端就可以访问其他网站。

图 7-11　测试

7.2　实训二　配置防火墙 DNAT

任务描述

某网络公司为了业务需求添加了一台 Web 服务器，允许外网用户通过合法 IP 地址访问 Web 服务器，网络管理员把此 Web 服务器连接到防火墙 E0/2 接口，把此接口加入防火墙 DMZ 区域，设置内网 Web 服务器的 IP 地址为 192.168.10.2/24，并利用公司外网 IP 地址 211.147.176.231/24 来发布内网 Web 服务器。

防火墙接口 IP 地址分配见表 7-2。

表 7-2　防火墙接口 IP 分配表

防火墙接口号	IP 地址
E0/0	192.168.1.1/24
E0/1	211.147.176.231/24
E0/2	192.168.10.1/24

网络拓扑图如图 7-12 所示。

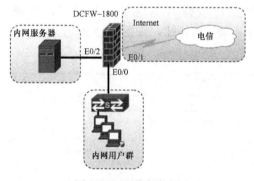

图 7-12　网络拓扑图

任务准备

DMZ：DMZ 是英文"demilitarized zone"的缩写，中文名称为"隔离区"，也称"非军事化区"。它是为了解决安装防火墙外部网络不能访问内部网络服务器的问题，而设立的一个非安全区域与安全区域之间的缓冲区，这个缓冲区位于企业内部网络和外部网络之间的小网络区域内，在这个小网络区域内可以放置一些必须公开的服务器设施，如企业Web 服务器等。

DNAT：DNAT（Destination Network Address Translation,目的地址转换）的作用是将一组本地内部 IP 地址映射到一组全球 IP 地址，这样外部客户就可以通过公网 IP 地址访问内部服务器，有效地解决了公网 IP 地址缺乏的问题。DNAT 是 NAT 的一种应用。

任务实施

1．对 IP 地址、区域、NAT、路由协议和防火墙策略等基础配置参照本章实训一。
2．选择"对象"选项卡中的"地址簿"，出现如图 7-13 所示界面。

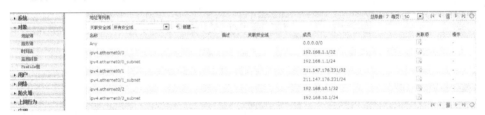

图 7-13　地址簿

3．单击"新建"，设置名称为"WEB"，并添加服务器 IP 地址为 192.168.10.2/32，如图 7-14 所示，单击"确认"按钮。

图 7-14　设置 Web 服务器 IP 地址

4．再次单击"新建"，设置名称为"外网 IP"，并添加外网接口 IP 地址为 211.147.176.231/32，单击"确认"按钮，如图 7-15 所示。

图 7-15　设置外网接口 IP 地址

5．在"防火墙"选项卡中的的"策略"菜单下单击"新建"，设置源安全域为"untrust"，源地址为"Any"，目的安全域为"dmz"，目的地址选择"WEB"，服务簿为"HTTP"，行为为"允许"，如图7-16所示，单击"确认"按钮。

图7-16 策略设置

6．选择"防火墙"选项卡中的"NAT"下的"目的NAT"，出现如图7-17所示界面。

图7-17 目的NAT列表

7．单击"新建"菜单中的"端口映射"，设置目的地址为"外网IP"，服务为"HTTP*"，映射到地址为"WEB"，映射到端口为"80"，如图7-18所示，单击"确认"按钮。

图7-18 目的NAT端口映射配置

8．外部客户机访问Web服务器测试。

图7-19 实训结果测试

7.3 实训三 配置防火墙 DHCP

任务描述

某网络公司遇到了一些员工不能上网的问题，起因是这些员工设置了错误的 IP 地址、网关和 DNS 服务器，网络管理员决定在防火墙上创建 DHCP 服务器，一次性解决这些问题，分配 IP 地址段为"192.168.1.10/24～192.168.1.250"，网关为 192.168.1.1，DNS 服务器为 8.8.8.8，并为经理建立了保留 IP 地址 192.168.1.66。

网络拓扑图如图 7-20 所示。

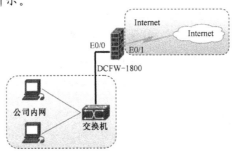

图 7-20 网络拓扑图

防火墙接口 IP 地址分配见表 7-3。

表 7-3 防火墙接口 IP 分配表

防火墙接口号	IP 地址
E0/0	192.168.1.1/24
E0/1	211.147.176.231/24

任务准备

DHCP：DHCP（Dynamic Host Configuration Protocol，动态主机配置协议）是一个局域网的网络协议，工作在 UDP 协议 68 和 69 端口，给内部网络或网络服务供应商自动分配 IP 地址。

任务实施

1. 对 IP 地址、区域、NAT、路由协议和防火墙策略等基础配置参照本章实训一。
2. 选择"网络"选项卡中的"DHCP"下的"地址池"，出现如图 7-21 所示界面。

图 7-21 DHCP 地址池列表

3．单击"新建"，设置地址池名为"地址池"，起始 IP 地址为"192.168.1.10"，结束 IP 地址为"192.168.1.250"，网关为"192.168.1.1"，网络掩码为"255.255.255.0"，租约为"1000000"秒，如图 7-22 所示，单击"确认"按钮。

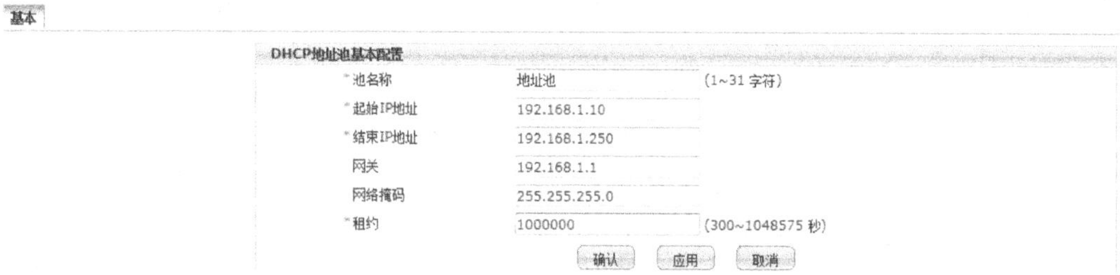

图 7-22　DHCP 地址池中设置的 IP 地址段

4．单击"修改"按钮，出现如图 7-23 所示界面。

图 7-23　DHCP 地址池中设置网关 IP 地址

5．选择"高级"选项卡，设置 DNS 为"8.8.8.8"，如图 7-24 所示，单击"应用"按钮。

图 7-24　DHCP 地址池中设置 DNS 服务器 IP 地址

6．选择"地址绑定"选项卡，单击"新建"，并设置 IP 为"192.168.1.66"，MAC 地址为经理 PC 机的 MAC 地址"10bf.4862.d665"，如图 7-25 所示，单击"确认"按钮。

图 7-25　设置保留的 IP 地址

7. 选择"网络"选项卡中的"DHCP"下的"服务",出现如图 7-26 所示界面。

图 7-26　DHCP 服务列表

8. 单击"新建",设置接口为内网接口"ethernet0/0",类型为"DHCP 服务器",地址池选择"地址池",如图 7-27 所示,单击"确认"按钮。

图 7-27　Ethernet0/0 接口启用 DHCP 服务器

9. 客户端测试,如图 7-28 所示。

图 7-28　实训结果测试

7.4　实训四　配置防火墙 URL 过滤

任务描述

某网络公司经理发现有部分员工在上班时间上网购物,主要集中在淘宝网,为了让员工安心工作,经理让网络管理员设置防火墙限制员工访问淘宝网站。

网络拓扑图如图 7-29 所示。

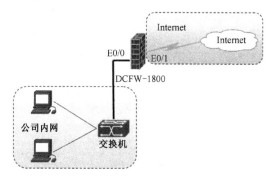

图 7-29　网络拓扑图

防火墙接口 IP 地址分配见表 7-4。

表 7-4　防火墙接口 IP 分配表

防火墙接口号	IP 地址
E0/0	192.168.1.1/24
E0/1	211.147.176.231/24

任务准备

URL（UniformResourceLocator，统一资源定位符）过滤：通过使用防火墙的 URL 过滤功能，设备可以控制用户的 PC 对某些网址的访问。URL 过滤功能包含以下组成部分：

- 黑名单：包含不可以访问的 URL。
- 白名单：包含允许访问的 URL。不同平台白名单包含的最大 URL 条数不同。
- 关键字列表：如果 URL 中包含有关键字列表中的关键字，则 PC 不可以访问该 URL。不同平台关键字列表包含的关键字条目数不同。
- 受限 IP：不受 URL 过滤配置影响，可以访问任何网站。
- 只允许用域名访问：如果开启该功能，用户只可以通过域名访问 Internet，IP 地址类型的 URL 将被拒绝访问。
- 只允许访问白名单里的 URL：如果开启该功能，用户只可以访问白名单中的 URL，其他地址都会被拒绝。

任务实施

1. 对 IP 地址、区域、NAT、路由协议和防火墙策略等基础配置参照本章实训一。
2. 选择"应用"选项卡中的"HTTP 控制"，出现如图 7-30 所示界面。

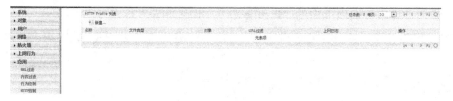

图 7-30　HTTP Profile 列表

3. 单击"新建",设置 Profile 名称为"URL",并启用"URL 过滤",如图 7-31 所示,单击"确认"按钮。

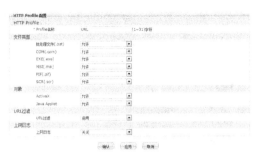

图 7-31 HTTP Profile 配置

4. 选择"应用"选项卡中的"URL 过滤",出现如图 7-32 所示界面。

图 7-32 URL 过滤配置

5. 在"黑名单"中添加淘宝网:www.taobao.com,如图 7-33 所示,单击"确认"按钮。

图 7-33 在黑名单中添加淘宝网

6. 选择"对象"选项卡中的"Profile 组",出现如图 7-34 所示界面。

图 7-34 Profile 组列表

7. 单击"新建",设置名称为"Profile",并添加组成员"URL(HTTP)",如图 7-35 所示,单击"确认"按钮。

图 7-35 添加 Profile 成员 URL 到 Profile 组中

8. 单击"防火墙"选项卡中的策略,出现如图 7-36 所示界面。

图 7-36 防火墙策略

9. 单击"新建",选择源安全域"trust"到目的安全域"untrust",服务簿改为"HTTP",行为为"允许",如图 7-37 所示,单击"确认"按钮建立一条允许内网访问外网 HTTP 服务的策略。

图 7-37 建立允许 HTTP 服务的策略

10. 选择此策略,单击"编辑",启用 Profile 组,并设置 Profile 组为"Profile",如图 7-38 所示,单击"确认"按钮。

图 7-38 策略的高级配置中启用 Profile 组

11. 在策略列表中将此策略移动到首位，如图 7-39 所示。

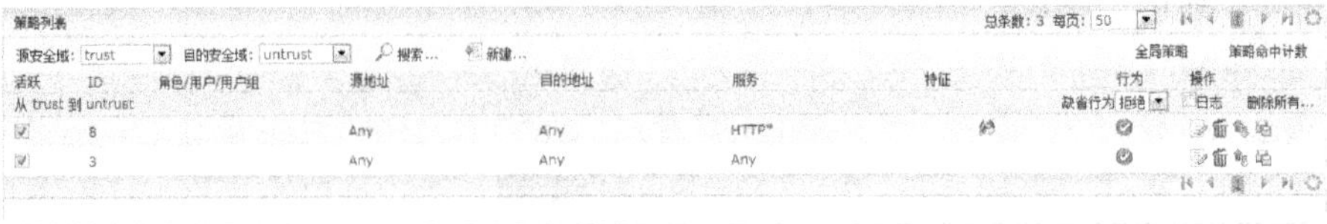

图 7-39　移动策略

12. 实验测试，内网用户在 IE 浏览器中输入 http://www.taobao.com，便会提示访问被拒绝，如图 7-40 所示。

图 7-40　拒绝访问淘宝

注意：在做实验过程中，如果学生先访问淘宝网站，然后按此实训步骤禁用淘宝网站，测试时访问淘宝成功，与实验预想不同。这是因为防火墙中对提前允许的数据流不再应用后设置的规则。应该是先访问其他网站"确认"能内网用户上网，然后禁用淘宝网站，最后测试访问淘宝。

7.5　实训五　配置防火墙网页内容过滤

任务描述

某网络公司经理发现上次的设置解决了员工上班时间访问淘宝的问题，但还是有很多人通过其他网站进行购物，所以这次要彻底封闭，经理让网络管理员设置防火墙限制员工访问带有"购物"字样的网站。

网络拓扑图如图 7-41 所示。

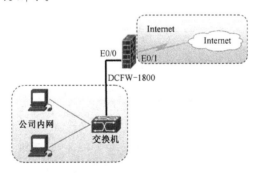

图 7-41　网络拓扑图

防火墙接口 IP 地址分配见表 7-5。

表 7-5 防火墙接口 IP 分配表

防火墙接口号	IP 地址
E0/0	192.168.1.1/24
E0/1	211.147.176.231/24

任务准备

网页内容过滤：网络内容过滤技术采取适当的技术措施，对互联网不良信息进行过滤，既可阻止不良信息对人们的侵害，同时，通过规范用户的上网行为，提高工作效率，合理利用网络资源，减少病毒对网络的侵害。

关键字：支持 UTF-8 和 GB18030 两种编码方式的关键字。关键字的表示形式可以为普通关键字和正则表达式。普通关键字按照字符进行逐字匹配，正则表达式按照正则表达式的计算结果进行匹配。DCFOS 支持 PCRE（Perl Compatible Regular Expressions）正则表达式语法。

类别：类别可包含多个关键字，通过类别名称来区分关键字。

类别组：类别组是类别的集合。

内容过滤 Profile：一个内容过滤 Profile 包含一个 HTTP 类别组。

任务实施

1. 对 IP 地址、区域、NAT、路由协议和防火墙策略等基础配置参照本章实训一。
2. 选择"应用"选项卡中的"内容过滤"下的"类别"，出现如图 7-42 所示界面。

图 7-42 类别输入

3. 单击"新建"，创建类别为"过滤"，如图 7-43 所示，单击"添加"按钮。

图 7-43 新建内容过滤类别

4. 选择"应用"选项卡中的"内容过滤"下的"关键字"，出现如图 7-44 所示界面。

图 7-44 输入关键字

5．单击"新建"，设置关键字为"购物"，类型选择"精确匹配"，类别选择"过滤"，如图 7-45 所示，单击"添加"按钮。

提示：添加或删除关键字之后，请单击"应用"按钮使其生效。

图 7-45　新建内容过滤关键字

6．选择"应用"选项卡中的"内容过滤"下的"类别组"，出现如图 7-46 所示界面。

图 7-46　类别组列表

7．单击"新建"，设置类别组为"过滤组"，如图 7-47 所示，单击"添加"按钮。

图 7-47　新建内容过滤类别组

8．单击新建内容过滤类别组"过滤组"菜单中的"编辑"按钮，出现如图 7-48 所示对话框。

图 7-48　类别组列表

9．单击"添加"，类别选择"过滤"，行为选择"拒绝"，如图 7-49 所示，单击"确认"按钮。

图 7-49　添加类别组成员

10．选择"应用"选项卡中的"内容过滤"下的"Profile"，出现如图 7-50 所示界面。

图 7-50　内容 Profile 列表

11．单击"新建"，设置 Profile 为"过滤 Profile"，类别组为"过滤组"，如图 7-51 所示，单击"确认"按钮。

图 7-51　新建内容 Profile

12．选择"对象"选项卡中的"Profile 组"，出现如图 7-52 所示界面。

图 7-52　Profile 组列表

13．单击"新建"，设置名称为"过滤 Profile 组"，并添加组成员"过滤 Profile"，如图 7-53 所示，单击"确认"按钮。

图 7-53　Profile 组配置

14．单击"防火墙"选项卡中的策略，出现如图 7-54 所示界面。

图 7-54　策略列表

15. 单击"新建",选择源安全域"trust"到目的安全域"untrust",服务簿改为"HTTP*",行为为"允许",如图 7-55 所示,单击"确认"按钮,建立一条允许内网访问外网 HTTP 服务的策略。

图 7-55　策略基本配置

16. 单击"编辑",启用 Profile 组,并设置 Profile 组为"过滤 Profile 组",如图 7-56 所示,单击"确认"按钮。

图 7-56　策略中启用 Profile 组

17. 将此策略移动到首位,如图 7-57 所示。

图 7-57　策略列表

18. 实验测试,内网用户在百度搜索引擎上输入购物,如图 7-58 所示。

图 7-58　访问百度页面

19. 单击"百度一下"按钮,就会出现"Internet Explorer 无法显示该网页"界面,如图 7-59 所示。

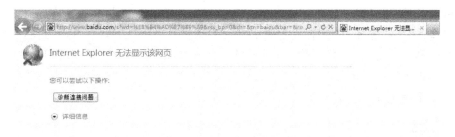

图 7-59　无法显示该网页

7.6　实训六　防火墙 Web 认证配置

任务描述

某网络公司需要对本公司员工的上网行为进行细化管理,决定在公司防火墙上启用 Web 认证,让公司员工上网时输入账户和密码,来保护公司网络。该网络公司内网网段为 "192.168.1.0/24"。

防火墙接口 IP 地址分配见表 7-6。

表 7-6　防火墙接口 IP 分配表

防火墙接口号	IP 地址
E0/0	192.168.1.1/24
E0/1	221.147.176.231/24

网络拓扑图如图 7-60 所示。

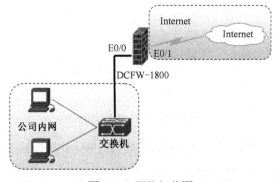

图 7-60　网络拓扑图

任务准备

Web 认证:Web 认证功能用来对通过设备访问 Internet 的用户进行身份认证。配置 Web 认证功能后,HTTP 请求会被重定向到 Web 认证登录页面,用户需要在该页面提供正确的用户名和密码；Web 认证成功后，系统会按照策略配置给 IP 地址分配角色，从而实现设备对不同用户的访问控制。

Web 认证策略规则：触发系统的 Web 认证功能，需要配置相应的策略规则。

重定向 URL：重定向 URL 功能是指用户在认证成功并返回认证页面后，弹出的新页面将会重定向到指定的 URL 页面。如果没有配置该功能，新弹出页面将返回用户输入的地址页面。该功能的正确执行需要浏览器关闭弹出窗口阻止程序。如果浏览器阻止弹出窗口，新弹出的页面将被阻止，需要手工确认才能打开。

任务实施

1. 对 IP 地址、区域、NAT、路由协议和防火墙策略等基础配置参照本章实训一。
2. 选择"网络"选项卡中的"Web 认证"，出现如图 7-61 所示界面。

图 7-61　Web 认证界面

3. 更改"模式"为"HTTP 模式"，启动 HTTP 认证，如图 7-62 所示，单击"确认"按钮。

图 7-62　启用 Web 认证

4. 选择"用户"选项卡中的"AAA 服务器"，出现如图 7-63 所示界面。

图 7-63　AAA 服务器列表

5. 单击"新建",设置名称为"aaa-Web",类型为"本地",如图 7-64 所示,单击"确认"按钮,建立名称为"aaa-Web"的 AAA 服务器。

图 7-64　新建 AAA 服务器

6. 选择"用户"选项卡中的"用户",出现如图 7-65 所示界面。

图 7-65　用户界面

7. 选择 AAA 服务器为"aaa-Web",选择"所有用户组",并单击"新建用户组",设置名称为"group",如图 7-66 所示,单击"确认"按钮。

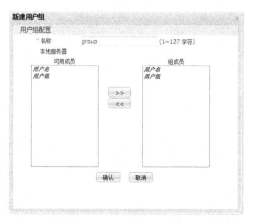

图 7-66　新建用户组 group

8. 设置 AAA 服务器为"aaa-Web",选择"所有用户",单击"新建用户",设置名称为"user",密码为"123123",如图 7-67 所示,单击"确认"按钮。

图 7-67　新建用户 user

9. 单击用户 user 的"编辑"按钮，将其所属组设置为"group"，如图 7-68 所示，单击"确认"按钮。

图 7-68　把用户加入组 group

10. 选择"用户"选项卡中的"角色"选项卡，出现如图 7-69 所示界面。

图 7-69　角色列表

11. 单击"新角色"，设置角色名称为"web"。描述为"允许访问 web"，如图 7-70 所示，单击"确认"按钮，建立名称为"web"的新角色。

图 7-70　新建角色

12. 单击"新角色映射"，角色映射名称为"role-map"，类型为"用户组"，用户组为"group"，角色为"web"，如图 7-71 所示，单击"添加"按钮，建立名称为 role-map 的角色映射。

图 7-71　新建角色映射

13．选择"用户"选项卡中的"AAA 服务器"，出现如图 7-72 所示界面。

图 7-72　AAA 服务器

14．单击"aaa-web 编辑"按钮，设置角色映射规则为"role-map"，如图 7-73 所示，单击"确认"按钮。

图 7-73　设置角色映射

15．选择"防火墙"选项卡中的"策略"，出现如图 7-74 所示界面。

图 7-74　防火墙策略列表

16．单击"新建"，设置源安全域为"trust"，源地址为"Any"，目的安全域为"untrust"，目的地址为"Any"，服务簿为"DNS"，行为为"允许"，单击"确认"按钮，建立允许内网用户访问外网 DNS 服务的策略，如图 7-75 所示。

图 7-75　建立 DNS 策略

17．再次单击"新建"，设置源安全域为"trust"，源地址为"Any"，目的安全域为"untrust"，目的地址为"Any"，服务簿为"Any"，行为为"Web 认证"，Web 认证为"aaa-web"，如

图 7-76 所示，单击"确认"按钮，建立一条允许内网用户访问外网所有服务需进行 Web 认证的策略。

图 7-76 对其他协议进行 Web 认证

18．单击上一步骤创建的策略"编辑"按钮，添加角色"UNKNOWN"，如图 7-77 所示，单击"确认"按钮，使本策略针对没有进行 Web 认证的角色或用户。

图 7-77 添加用户 UNKNOWN

19．再次单击"新建"，设置源安全域为"trust"，源地址为"Any"，目的安全域为"untrust"，目的地址为"Any"，服务簿为"HTTP"，行为为"允许"，如图 7-78 所示，单击"确认"按钮，建立一条允许内网用户访问外网 HTTP 服务的策略。

图 7-78 建立允许 HTTP 规则

20. 单击上一步骤创建的策略"编辑"按钮，添加角色"web"，如图 7-79 所示，单击"确认"按钮，使本策略仅允许内网角色"web"访问外网 HTTP 服务。

图 7-79　允许"web"角色上网

21. 任务测试，内网用户打开 IE 浏览器后输入某网站，就会重定向到认证登录界面，输入用户名"user"和密码"123123"，如图 7-80 所示。

图 7-80　Web 认证登录系统

22. 单击"登录"后，出现如图 7-81 所示界面。这样经过认证的用户，上网期间保持此页面打开，才能访问互联网。

图 7-81　认证通过界面

7.7 实训七 防火墙 IP-MAC 绑定配置

任务描述

某网络公司管理员发现最近 ARP 病毒肆虐，造成某些时间段上网困难，为了解决这一问题，管理员决定在公司内部对 IP 地址和 MAC 地址进行绑定，为了节约时间采用动态绑定的方式。公司内网网段为 "192.168.1.0/24"。

防火墙接口 IP 地址分配见表 7-7。

表 7-7 防火墙接口 IP 分配表

防火墙接口号	IP 地址
E0/0	192.168.1.1/24
E0/1	221.147.176.231/24

网络拓扑图如图 7-82 所示

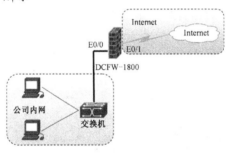

图 7-82 网络拓扑图

任务准备

IP-MAC 绑定：为加强网络安全控制，最有效的解决方法就是把 IP 地址和 MAC 绑定，从物理通道上隔离了盗用者。

动态 IP-MAC 绑定：通过 ARP 学习功能、ARP 扫描功能以及 MAC 学习功能获得绑定信息为动态绑定信息。

静态 IP-MAC 绑定：手工配置的 IP-MAC 绑定信息。

任务实施

1. 对 IP 地址、区域、NAT、路由协议和防火墙策略等基础配置参照本章实训一。
2. 选择"防火墙"选项卡中的"二层防护"菜单下的"静态绑定"，出现如图 7-83 所示界面。

图 7-83 IP-MAC 列表

3．单击"扫描 IP-MAC"，设置起始 IP 地址为"192.168.1.1"，终止 IP 地址为"192.168.1.254"，如图 7-84 所示，单击"确认"按钮。

图 7-84　扫描 IP-MAC

4．单击"绑定"，选择"绑定所有 IP-MAC"，如图 7-85 所示，单击"确认"按钮。这样绑定 IP-MAC 地址的用户就不可以自己随便更改 IP 地址，如果改成其他 IP 地址则不能访问网络。

图 7-85　绑定 IP-MAC

5．选择"防火墙"选项卡中的"二层防护"菜单下的"静态绑定"，出现如图 7-86 所示界面，显示成功绑定 IP-MAC。

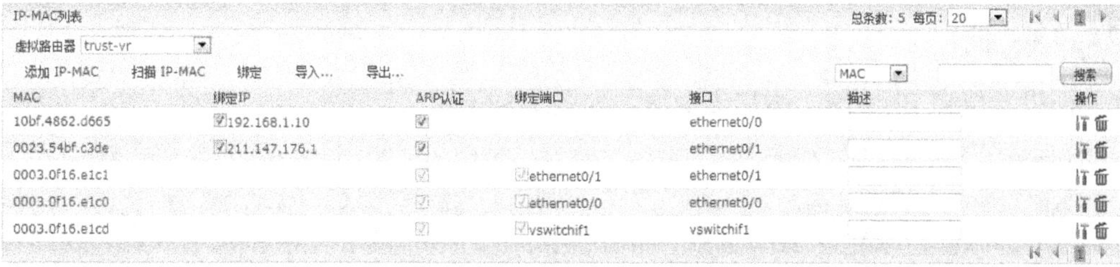

图 7-86　绑定界面

7.8　实训八　防火墙负载均衡配置

任务描述

某网络公司现在业务极度依赖于网络，为了增加带宽和防止上网单点失败的问题，所以申请了联通和电信各一条线路，其中电信 IP 地址"211.147.176.231/24"，网关为"211.147.176.1"，另外联通 IP 地址"222.173.165.156/24"，网关为"222.173.165.1"，并在防火墙中针对此两条线路启用负载均衡。

防火墙接口 IP 地址分配见表 7-8。

表 7-8 防火墙接口 IP 分配表

防火墙接口号	IP 地址
E0/0	192.168.1.1/24
E0/1	221.147.176.231/24
E0/2	222.173.165.156/24

网络拓扑图如图 7-87 所示。

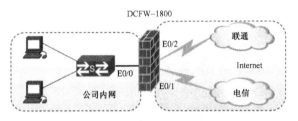

图 7-87 网络拓扑图

任务准备

负载均衡（Load Balance）：负载均衡就是将负载（工作任务）进行平衡、分摊到多个操作单元上，例如网络多个出接口、Web 服务器、FTP 服务器、企业关键应用服务器和其他关键任务服务器等，从而共同完成工作任务。负载均衡是对网络的性能优化。对于网络应用而言，并不是一开始就需要负载均衡。当网络应用的访问量不断增长，单个处理单元无法满足负载需求，网络应用流量将要出现瓶颈时，负载均衡才会起到作用。

任务实施

1. 设置接口 IP 地址，添加安全域，如图 7-88 所示。

图 7-88 网络接口列表

2. 选择"网络"选项卡中的的"路由"菜单下的"目的路由"，出现如图 7-89 所示界面。

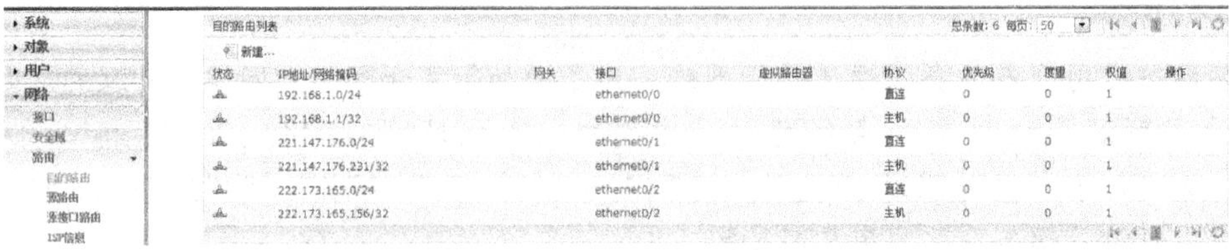

图 7-89 目的路由列表

3. 单击"新建",目的 IP 为"0.0.0.0",子网掩码为"0.0.0.0",下一跳为"网关",网关为"211.147.176.1",优先级为"1",路由权限为"1",如图 7-90 所示,单击"确认"按钮。

4. 再次单击"新建",目的 IP 为"0.0.0.0",子网掩码为"0.0.0.0",下一跳为"网关",网关为"222.173.165.1",优先级为"1",路由权限为"1",如图 7-91 所示,单击"确认"按钮。

图 7-90　创建到电信默认路由　　　　图 7-91　创建到联通默认路由

5. 设置防火墙负载均衡方式,必须在命令行下进行,首先通过 Console 口连接防火墙,并输入防火墙用户名 admin 和密码 admin。

```
DCFW-1800# config              // 进入全局配置模式
DCFW-1800(config)# ecmp-route-select ?        //显示防火墙负载均衡方式
  by-5-tuple        Configure ECMP Hash As 5 Tuple
  by-src            Configure ECMP Hash As Source IP
  by-src-and-dst    Configure ECMP Hash As Source IP and Dest IP
DCFW-1800(config)# ecmp-route-select by-5-tuple      //选择"by-5-tuple",
也就是五元组方式(源 IP 地址、目的 IP 地址、源 MAC 地址、目的 MAC 地址和协议号)
```

6. 设置监控地址(网关)。

```
DCFW-1800(config)# track "e0/1"        //创建监控地址名
DCFW-1800(config-trackip)# ip 211.147.176.1 interface ethernet0/1
//设置监控地址为"211.147.176.1"
DCFW-1800(config-trackip)# exit        //退出
DCFW-1800(config)# track "e0/2"        //创建监控地址名
DCFW-1800(config-trackip)# ip 222.173.165.1 interface ethernet0/2
//设置监控地址为"222.173.165.1"
DCFW-1800(config-trackip)# exit        //退出
```

7. 在外网接口应用监控地址。

```
DCFW-1800(config)# interface e0/1           //进入 E0/1 接口
DCFW-1800(config-if-eth0/1)# monitor track e0/1      //引用监控地址
DCFW-1800(config-if-eth0/1)# exit           //退出
DCFW-1800(config)# interface e0/2           //进入 E0/2 接口
DCFW-1800(config-if-eth0/2)# monitor track e0/2      //引用监控地址
DCFW-1800(config-if-eth0/1)# exit           //退出
```

7.9 实训九 防火墙禁用实时通信类工具配置

任务描述

某网络公司经理发现最近有部分员工经常上 QQ 和 MSN 等聊天工具，影响了公司正常的业务，并且公司业务也不需要通过此类即时通信软件，所以决定通过防火墙禁用 QQ 和 MSN。

该网络公司内网网段"192.168.1.0/24"，防火墙接口 IP 地址分配见表 7-9。

表 7-9 防火墙接口 IP 分配表

防火墙接口号	IP 地址
E0/0	192.168.1.1/24
E0/1	221.147.176.231/24

网络拓扑图如图 7-92 所示。

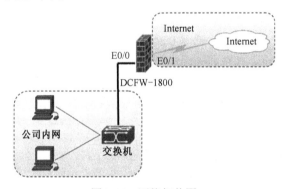

图 7-92 网络拓扑图

任务准备

IM：即时通信（Instant Messenger，IM）是一种基于互联网的即时交流消息的业务，例如 MSN、QQ、UC、YY 等。

任务实施

1. 对 IP 地址、区域、NAT、路由协议和防火墙策略等基础配置参照本章实训一。
2. 选择"网络"选项卡中的"安全域"，出现如图 7-93 所示界面。

图 7-93 安全域界面

3. 单击 untrust 区域 "编辑" 按钮，启用应用识别功能，如图 7-94 所示，单击 "确认" 按钮。

图 7-94　启用应用识别

4. 选择 "防火墙" 选项卡中的的 "策略"，出现如图 7-95 所示界面。

图 7-95　防火墙策略界面

5. 单击 "新建"，设置源安全域为 "trust"，目的安全域为 "untrust"，服务簿为 "QQ*"，行为为 "拒绝"，如图 7-96 所示，单击 "确认" 按钮。

图 7-96　拒绝 QQ 策略

6. 将此 "策略" 移动到首位，如图 7-97 所示。

图 7-97　策略应用顺序调整

7. 再次单击 "新建"，设置源安全域为 "trust"，目的安全域为 "untrust"，服务为 "MSN*"，行为为 "拒绝"，如图 7-98 所示，单击 "确认" 按钮。

图 7-98　拒绝 MSN 策略

8. 将此"策略"移动到首位，如图 7-99 所示。

图 7-99　策略应用顺序调整

9. 用户测试，如图 7-100 所示。

图 7-100　QQ 登录超时界面

7.10　实训十　防火墙应用 QOS 配置

任务描述

某网络公司出口带宽 50Mbps，网络管理员发现最近有用户提出不能正常上网，经过数据包检测后发现 P2P 应用占用了大量带宽，决定限制 P2P 应用下行带宽为 10M，上传最大 5M，HTTP 和 SMTP 应用下载保障 20M，上传保障 10M。

防火墙接口 IP 地址分配见表 7-10。

表 7-10　防火墙接口 IP 分配表

防火墙接口号	IP 地址
E0/0	192.168.1.1/24
E0/1	211.147.176.231/24

网络拓扑图如图 7-101 所示。

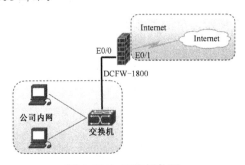

图 7-101　网络拓扑图

防火墙的具体应用 项目 7

任务准备

QoS：QoS（Quality of Service）即"服务质量"。它是指网络为特定流量提供更高优先服务的同时控制抖动和延迟的能力，并且能够降低数据传输丢包率。当网络过载或拥塞时，QoS能够确保重要业务流量的正常传输。

任务实施

1. 对IP地址、区域、NAT、路由协议和防火墙策略等基础配置参照本章实训一。
2. 选择"QoS"选项卡中的"接口带宽"，出现如图7-102所示界面。

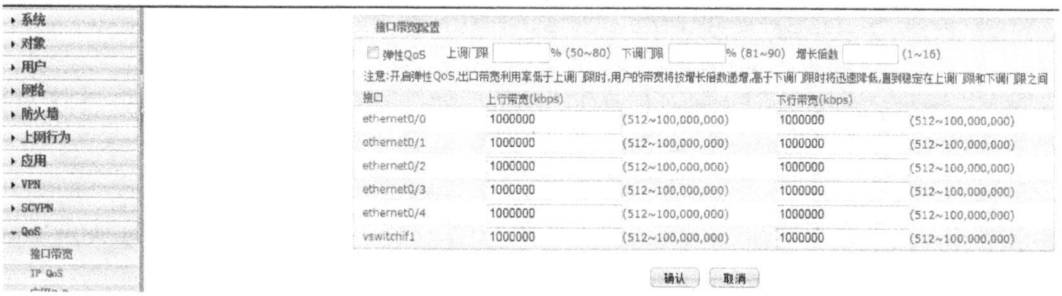

图7-102 接口带宽

3. 配置"ethernet0/1"接口，上行带宽为"50000kbps"，下行带宽为"50000kbps"，启用"弹性QoS"，如图7-103所示，单击"确认"按钮。

图7-103 配置接口带宽

4. 选择"网络"选项卡中的"安全域"，出现如图7-104所示界面。

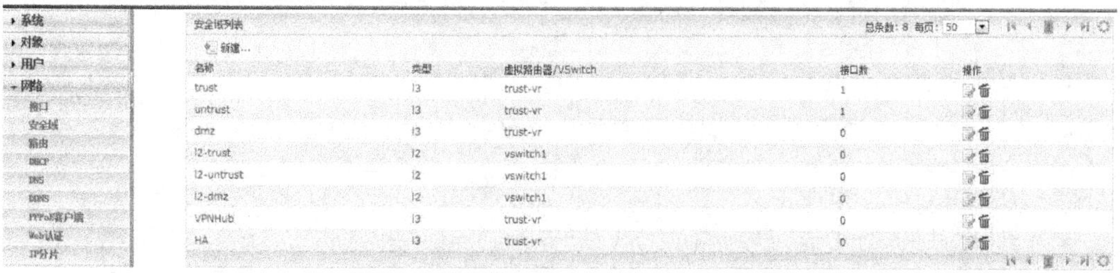

图7-104 安全域界面

5. 单击"untrust 修改"按钮,启用"应用识别"如图 7-105 所示,单击"确认"按钮。

图 7-105　启用应用识别

6. 选择"QoS"选项卡中的"应用 QoS",出现如图 7-106 所示界面。

图 7-106　应用 QoS

7. 设置规则名称为"P2P",接口绑定为外网接口"ethernet0/1",添加服务成员"P2P 下载",宽带上行最大带宽为"5000",宽带下行最大带宽为"10000",如图 7-107 所示,单击"添加"按钮。

图 7-107　设置 P2P 带宽

8. 设置规则名称为"ensure1",接口绑定为外网接口"ethernet0/1",添加服务成员 "HTTP 和 SMTP",宽带上行最小带宽为"10000",如图 7-108 所示,单击"添加"按钮。

图 7-108　设置 HTTP 和 SMTP 上行带宽

9. 设置规则名称为"ensure2",绑定接口为外网接口"etherne0/1",添加服务成员"HTTP 和 SMTP",宽带下行最大带宽为"20000",如图 7-109 所示,单击"添加"按钮。

图 7-109　设置 HTTP 和 SMTP 下行带宽

7.11　实训十一　配置防火墙 SSL VPN

任务描述

某网络公司有一些员工需要经常出差,并且在出差时需要访问公司内网服务器,管理员为了保证数据在传输过程中安全传输,并且不在客户端安装额外软件,应用 IE 就可以拨入公司内网,决定在防火墙中启用 SSL VPN,为远程客户机分配 IP 地址段为"192.168.2.0/24"。

防火墙接口 IP 分配见表 7-11。

表 7-11　防火墙接口 IP 分配表

防火墙接口 IP	IP 地址
E0/0	192.168.1.1/24
E0/1	211.147.176.231/24
tunnel1	192.168.2.1/24

网络拓扑图如图 7-110 所示。

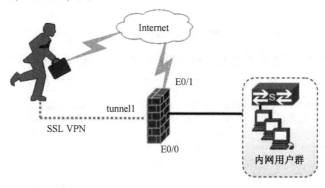

图 7-110　网络拓扑图

任务准备

SSL VPN： SSL（Secure Sockets Layer）是由 Netscape 公司开发的一套 Internet 数据安全协议，当前版本为 3.0。它已被广泛地用于 Web 浏览器与服务器之间的身份认证和加密数据传输。SSL 协议位于 TCP/IP 协议与各种应用层协议之间，为数据通信提供安全支持。SSL VPN 是解决远程用户访问敏感公司数据最简单、最安全的解决技术。与复杂的 IPSec VPN 相比，SSL 通过简单易用的方法实现信息远程连通。任何安装浏览器的机器都可以使用 SSL VPN，这是因为 SSL 内嵌在浏览器中，它不需要像传统 IPSec VPN 一样必须为每一台客户机安装客户端软件。

任务实施

1. 选择"SCVPN"选项卡中的"地址池"，出现如图 7-111 所示界面。

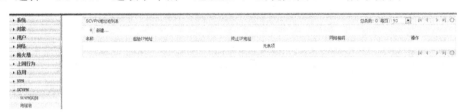

图 7-111　SCVPN 地址池

2. 单击"新建"，设置地址池名为"SCVPN"，起始 IP 地址为"192.168.2.10"，终止 IP 地址为"192.168.2.50"，网络掩码为"255.255.255.0"，如图 7-112 所示，单击"确认"按钮。

图 7-112　新建地址池

3. 选择"SCVPN"选项卡中的"SCVPN 实例"，出现如图 7-113 所示界面。

图 7-113　SCVPN 实例界面

4．单击"新建"，设置名称为"SCVPN"，单击"确认"按钮，如图 7-114 所示。

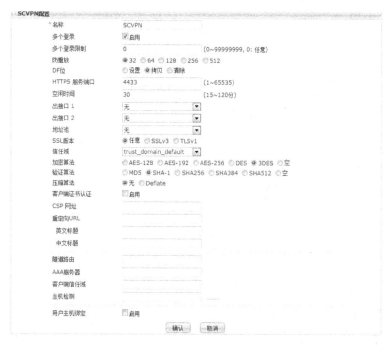

图 7-114　设置 SCVPN 实例

5．单击"编辑"按钮，设置出接口为"ethernet0/1"，地址池为"SCVPN"，隧道路由添加"192.168.1.0/24"，AAA 服务器添加"local"，单击"确认"按钮。

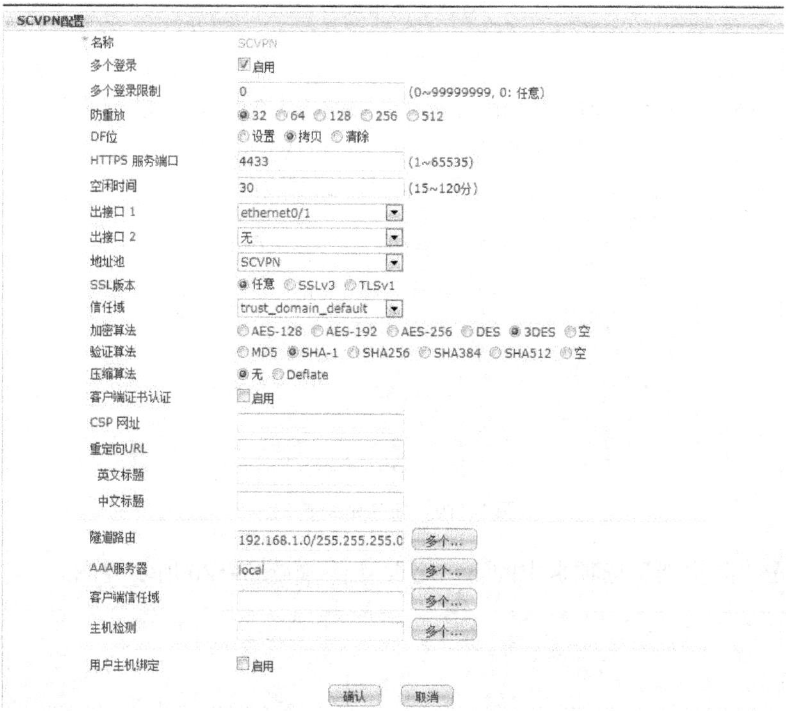

图 7-115　编辑 SCVPN 实例

6．选择"网络"选项卡中的"安全域"，出现如图7-116所示界面。

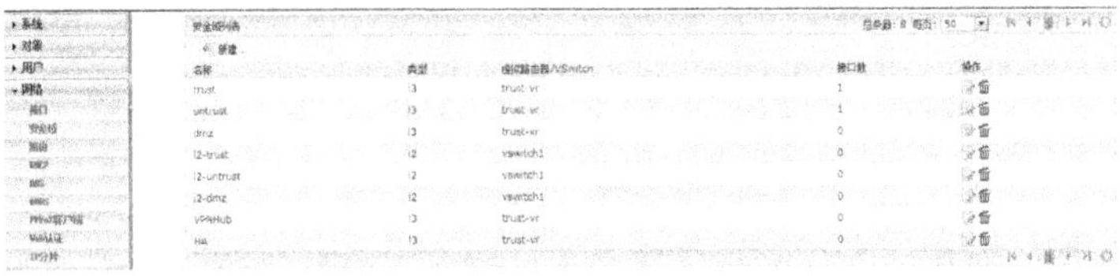

图7-116 安全域界面

7．单击"新建"，并设置安全域名称为"SCVPN"，安全域类型为"三层安全域"，如图7-117所示，单击"确认"按钮。

图7-117 添加安全域

8．选择"网络"选项卡中的"接口"，出现如图7-118所示界面。

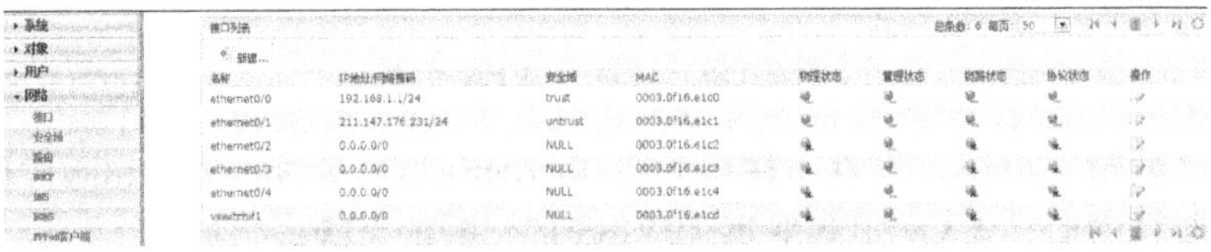

图7-118 网络接口界面

9．单击"新建"菜单下的"隧道接口"，设置接口名为"tunnel1"，安全域类型为"三层安全域"，安全域为"SCVPN"，IP地址为"192.168.2.1/24"，隧道类型为"SCVPN"，VPN名称为"SCVPN"，如图7-119所示，单击"确认"按钮。

图7-119 新建隧道接口

10．单击"防火墙"选项卡中的"策略"，出现如图7-120所示界面。

图7-120 防火墙策略界面

11. 单击"新建",设置源安全域为"SCVPN",目的安全域为"trust",服务簿为"Any",行为为"允许",如图 7-121 所示,单击"确认"按钮。

图 7-121　新建策略

12. 单击"用户"选项卡中的"用户",出现如图 7-122 所示界面。

图 7-122　用户界面

13. 单击"新建用户",设置名称和密码,如图 7-123 所示,单击"确认"按钮。

图 7-123　新建用户

14. 在客户机上打开浏览器,在地址栏中输入:"https://211.147.176.231:4433",输入用户名和密码,如图 7-124 所示,单击"确认"按钮。

图 7-124　公网客户机拨入界面示意图

15. 进入如图 7-125 所示界面，单击下载 SSL VPN 客户端。

图 7-125　弹出 DigitalChina Secure Connect 控件

16. 下载安装"DigitalChina Secure Connect"控件完成后，就会在右下角的任务栏中出现拨号成功的标识，如图 7-126 所示。

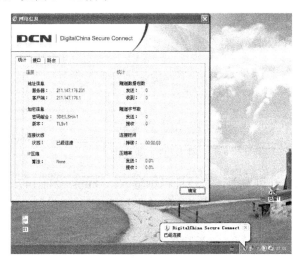

图 7-126　实训结果测试